T0276070

Cold Air Accumulation and the Grower's Guide to Frost Protection

Steve Hammersmith

WestBow Press books may be ordered through booksellers or by contacting:

WestBow Press
A Division of Thomas Nelson & Zondervan
1663 Liberty Drive
Bloomington, IN 47403
www.westbowpress.com
1 (866) 928-1240

ISBN: 978-1-4908-4927-0 (sc)
ISBN: 978-1-4908-4928-7 (e)

Library of Congress Control Number: 2014916302

Printed in the United States of America.

WestBow Press rev. date: 11/10/2014

WestBow
PRESS
A DIVISION OF THOMAS NELSON
& ZONDERVAN

Table of contents

Introduction 3

Chapter 1- Analysis and Evaluation techniques.................... 5

Chapter 2- Types of Frost Events 8

    Advection Freezes – regional frost events 8
    Radiation Frost – micro climate frost events 10
    Factors that need to be present to have a
    Radiation frost event 11

Chapter 3- Building Blocks of Frost 14

    Inversion layer 14
    Thermal gradient 16
    Atmospheric stratification and
    thermal boundary lines 16
    Down slope / katabatic wind 19
    Return period of a frost event 21
    Frost damage vs. the potential for
    frost damage and super cooling 23

Chapter 4- Myths of Frost Protection 25

Chapter 5- Frost Protection Tools 32

    Frost protection tools- active measures 32

        Overhead sprinklers 32
        Under tree sprinklers and flood irrigation.......... 35
        Wind machines 37
        Shur Farms Cold Air Drain® 40
        Heaters 43
        Fog 46
        Helicopters 47
        Chemical sprays 48
        Burning debris 49

Frost protection tools – passive measures   49

    Site and varieties selection ....…...   49
    Barriers and diversions   50
    Soil management   52
    Cover Crop management   53
    Late pruning, chemical and other
    Dormancy extending methods...................….....   54
    Methods of frost protection that should
    never be used together ....…..…   54
    Methods of frost protection that should
    not be used in certain situations ....…...….......   55

Chapter 6 – Cold air Accumulation   57

    Patterns and analysis   57
    Static and dynamic accumulation   59
    Accumulation process # 1 – Cold air lake   59
    Accumulation process # 2 – Cold air rivers ..…...........   62
    Accumulation process # 3 – Converging air currents ...   64
    Accumulation process # 4 – Lessening slope   66
    Accumulation process # 5 – Plain slope   69
    Accumulation process # 6 – Flooding   71
    Accumulation process # 7 – Flat area   74

Chapter 7 – Information Gathering and sample evaluations   75

    Appendix 1 – Evaluation and information form   75
    Appendix 2 – Summary frost analysis   78
    Appendix 3 – Optimization plan and summary analysis.   82

SHuR FARMS®
Frost Protection

Division Recovery P.T. Corporation
Toll Free 877-842-9688
1890 N. 8th Street, Colton, CA 92324   (909) 825-2035, (909) 825-2611 Fax
info@shurfarms.com
www.shurfarms.com

COLD AIR ACCUMULATION AND THE GROWER'S GUIDE TO FROST PROTECTION

Introduction

The occurrence of frost is a serious problem in some areas. Where vineyards and tree fruit crops are planted in rolling hills and surrounded by lush natural vegetation, the views and esthetics may be prized almost as much as the crops themselves. Environmental considerations of air and water pollution, noise and sight pollution and other undesired side effects of frost protection methods should be considered before deciding on a strategic plan.

In the last 20 years there have been significant advances in the ability to treat chronic frost problems. The proliferation of geospatial data sources and the availability of GIS, SRTM, DEM and other topographic and 3D visualization tools are helping to provide new understanding of the forces that cause frost damage. These forces are consistent and predicable, but the impact of them is changing. As farms are forced into less desirable areas because of urbanization and climate change is causing earlier bud breaks or changing weather necessitates different varieties or crops there will be more need to understand and mitigate these forces.[1]

---

[1] Responses of Spring Phenology to Climate Change. Franz-W. Badeck, Alberte Bondeau, Kristin Bottcher, Daniel Doktor, Wolfgang Lucht, Jorg Schaber and Stephen Sitch*New Phytologist*, Vol. 162, No. 2 (May, 2004), pp. 295-309
(article consists of 15 pages)Published by: Blackwell Publishing on behalf of the New Phytologist Trust   Stable URL: http://www.jstor.org/stable/1514502

The old methods of frost protection were never very effective and certainly not efficient. It is not sustainable any longer to burn unlimited amounts of fossil fuels or cause agonizing levels of pollution for dubious protection.

It is undisputable that most frost damage occurs in areas where there is cold air accumulation. With this book it is my hope that you gain a new understanding of the causes of this accumulation and be better able to identify and remedy the problem.[2]

---

[2] e.g., Whiteman and McKee, 1982; Kondo et al., 1989; Kondo and Okusa, 1990; Whiteman, 1990; O'Steen, 2000; Clements et al., 2003; Clements and Zhong, 2004; Zangl, 2005; Steinacker et al., 2007]

Department of Geography, The Australian National University, Canberra, Australia, 2 CSIRO Division of Water Resources, Canberra, Australia
Frost Risk Mapping for Landscape Planning: A Methodology
G. P. Laughlin ~ and J. D. Kalma 2
With 8 Figures
Received March 25, 1988
Revised September 21, 1989

Note: Shur Farms Frost Protection® identification of cold air accumulation processes is outlined in chapter 6

CHAPTER 1 – Analysis and Evaluation

Frost protection is a method of risk management. It should be viewed with the idea that no system can be put into place that will protect 100% every year, but overall there must be a satisfactory return on investment. Any decision on the acquisition of frost protection equipment should consider the economics, as well as the environment and esthetics. The factors that will need to be considered are the crop value, the risk of frost damage, operating costs including man hours, fuel and maintenance, and the original acquisition cost of the equipment used. The effectiveness and return on investment of a frost protection system considers the frost losses over a period of years in the protected area compared to the same area for the same period of time without protection. It is not feasible to fully understand the impact of a frost protection system in the first year of operation.

To design a system, a two part <u>Summary Frost Risk Analysis</u> of the area is performed. The first part will consist of relevant information including frost history, orchard specifics (acreage, age, skirt/cordon height, etc.) and the frost history of the surrounding area. If reliable information is available from nearby orchards and then compared to the subject area, the analysis can give a feel for the risk and probable intensity of frost damage, even when there is little actual information from the protected area itself. The second part of the analysis consists of topographic and air flow models and will identify where cold air is entering, exiting and accumulating in the growing area. The designed frost protection system will optimize the growing area with passive measures that naturally enhance cold air drainage and block or retard cold air intrusion. Active frost protections will then forcibly remove or mitigate the effects of any remaining cold air accumulation in the growing area. The analysis will identify the appropriate primary and, if needed, secondary active frost protection methods that can be used. The analysis will detail the passive methods that enhance the active methods used. An example Summary Analysis is provided in Appendix 1 - 3.

To protect new plantings or plantings without a frost damage history a <u>Preventative Frost Protection</u> system is designed. This design is based on either a Summary Frost Risk Analysis or a Full Frost Risk Analysis. There are downsides to a preventative system. Some methods of frost protection can cause damage to the crops if not used properly where none may have occurred without it, and the cost of running frost protection will add up as an expense that might not have been needed. Methods that can cause damage or have high acquisition or operating costs should be avoided.

A Preventative Frost Protection system designed on the basis of a Summary Frost Risk Analysis assumes a moderate to severe frost risk in the areas of cold air accumulation. In reality, there may or may not be a frost risk in these areas because cold air accumulation alone does not assure that the area will fall below the critical temperature for the crop that is planted there, only that it will be colder than other areas that adequately drain themselves. An accumulation area where peaches are grown may have substantial frost damage, but if lemons are planted in the same area, there may be no frost damage. A frost area is only a frost area if the crop that is planted there is susceptible to the minimum temperatures at the critical time.

A Preventative Frost Protection system based on a Full Frost Risk Analysis uses the data collected to create a 'virtual' frost damage history. A virtual damage history can be compiled for any area, even if it is newly graded or the outside factors contributing to the micro climate inside the protected area have recently changed. If the outside factors change after the data is collected, such as a tree line removed or a structure is built, the acquired data may no longer be relevant.

A Full Frost Risk Analysis is made by installing data loggers in strategic spots within the growing area and an additional data logger in a control area. A local regional weather station (CIMIS, Adcon, PAWS, etc.) is used as the base. The control data and the base data must come from the same region as the growing area because micro climates within a regional climate will have consistent temperature differences, but there is no correlation of temperature differences between micro climates in different regional climates.[3]

Data logger locations are determined by identifying the accumulation areas and cold air streams of the new growing area with a topographic and air flow analysis. By placing the loggers in areas presumed to be colder (suspected accumulation areas) and areas presumed to be warmer (hillsides and other well drained areas), a micro climate map can be produced showing the relative temperatures in the growing area. Relative temperature data will confirm the existence of cold air accumulation areas and drainage areas.

The readings from the data loggers will give temperature differences within the orchard as well as the differences with the control and weather station. If the weather station has several years of history, then conclusions as to the actual temperature within the growing area during past radiation frost events can be made by correlating the temperature of the weather station with the data logger at a specific site and then correcting for the differential. For instance, if a data logger shows a consistent maximum of -3 deg. F in relation to the weather station during a radiation event, then whatever temperature that is recorded on the weather station during the growing season is presumed to be 3 deg. higher than that spot during a radiation event. When this information from prior years is recorded on a spread sheet, a virtual frost history can be compiled on newly planted orchards. This history can include actual low temperatures, dates

---

[3] **THE RELATIONSHIP BETWEEN THE** MACRO- AND MICROCLIMATE t
R. M. HOLMES AND A. NELSON DINGLE
Agrometeorological Section, Plant Research Institute, Research Branch, Department of Agriculture, Ottawa, Ont. (Canada)
Department of Meteorology and Oceanography, The University of Michigan, Ann Arbor, Mich. (U.S.A.)
(Received April 30, 1964)

and frequency of frost events. With this information the grower can match the specific requirements of each variety of crops with the most acceptable area within the orchard and determine the amount of frost protection that will be needed.

An <u>Executive Style Risk Analysis</u> is basically an abbreviated frost risk analysis. It can be made by installing data loggers in strategic spots within the growing area as in a full frost risk analysis, and a control data logger in a known area that is growing the same variety of crops as what is to be planted. Instead of comparing temperatures to a weather station and creating a virtual frost history, by choosing a control area that has an acceptable frost damage history, temperature comparisons can be made and the grower can assess the viability of the new planting. If the temperatures are the same in the proposed area as in the known area, then the frost risk is the same. If the temperatures are less, the risk is higher.

Data logger data must be compiled during radiation events. The most beneficial data is from the long uninterrupted nights. Several months of data may need to be considered in order to get a relevant and useable sample. This is time consuming and tedious work and has a high engineering cost. Due to the high cost of a full Frost Risk Analysis, it is often more economical to perform the Executive Risk Analysis or to simply assume moderate to severe frost risk in the accumulation areas and design and install a preventative frost protection system accordingly.

The first step in implementing a frost protection system, whether this is a new planting or an existing orchard, is to optimize the situation with passive measures. What this entails is to identify the influx of cold air and block or divert its incursion into the growing area, enhance the natural drainage capacity, plan cordon or skirt heights in accordance with slope angles, plan and manage cover crops, and plan and manage natural vegetation barriers.

If an optimization plan is developed prior to planting, then even the row directions and locations of canals, soil berms, and buildings can be optimized. The optimization plan is based on the site analysis.

Chapter 2:   Types of Frost events

There are two types of frost conditions; Advection freeze and Radiation frost. These two events are quite different in character and physical properties. Some of the major differences are:

1.   During an Advection freeze, there is little or no atmospheric stratification and so there is no inversion layer; Radiation frosts are characterized by moderate to strong inversion and stratified air layers.
2.   Cold air in an advection freeze comes from the sky, usually as a winter storm; Radiation frosts are driven by heat loss from the ground, which cools the ground and then the ground cools the air.
3.   Advection freeze can last throughout the night and daytime for prolonged periods of time; Radiation frosts and atmospheric stratification mainly occur at night.
4.   Advection freeze can be accompanied by cloudy conditions and wind, rain, snow, or hail; Radiation frost occurs under no clouds and no wind conditions.
5.   Advection freeze is a regional climactic event and will affect a large area; Radiation frost is a micro climate event and while there is a correlation between temperatures in different areas in the same region, there is no connection between one micro climate and the next.

Advection Freezes- Regional frost events[4]

Advection freezes are meteorological events. They happen when winter storm conditions usually originating from northern regions many miles away move into an area as a weather system. The sky can be cloudy and often there is wind, rain, hail, or snow. During these winter conditions, except in cases of crops that bud year round, most trees and vineyards are still safely dormant. The danger in advection freezes is when they come later than normal. The actual temperatures that accompany advection freezes are not normally unusually cold, but rather they are unusually late. Sometimes they come only a week or two later than normal, allowing for the trees to come out of dormancy and be at risk for frost damage.

Advection freezes can be mixed with elements of radiation frost when a clear, cold mass of air moves into an area. This mass of cold winter air may come in over the ground at varying

---

[4] Published by the Program of Viticulture and Enology. Department of Horticulture. Michigan State University. 2003. http://docs.google.com/gview?a=v&q=cache:iLMXoY0P8_EJ:www.grapes.msu.edu/pdf/cultural/factors-related.pdf+factors+influencing+spring+freeze+damage+to+developing+grape+shoots&hl=en&gl=us&sig=AFQjCNHcZ1HZJ04ZFArg7znnS66kMqrPNQ (generally describing advection).

Frost Damage and Management in New Zealand Vineyards© Trought, Howell and Cherry: Lincoln University.

heights. Because the cold air moves in a mass over the ground this is a weather event and these events can be classified as advection even though there are the main elements of radiation frost present- clear skies and no wind. Sometimes, these cold air masses come in above the ground. There may be a mass of air, for example, that is 200 feet thick, coming in at 100 - 300ft above the ground. In this case a grower that is lower than 100 ft from the regional floor might not be affected, but the growers between 100 and 300 ft will suffer frost damage. In addition to this inconsistency, the mass of air can move with an irregular leading edge, leaving areas several miles inland with more damage than some areas closer to the source. It is precisely these irregularities that make these freezes almost impossible to protect against.

The good news is that, advection freezes during the critical growing times of budding and pollination are relatively rare. The main reason for the rarity of these events is that most growers will not intentionally plant crops that are not suitable for their regional climate conditions, and advection freezes are 'regional' frost events. A crop that regularly comes out of dormancy prior to the last winter storm is not suitable to the regional climate. If a regional frost event occurs more often than what is acceptable for the economic well being of the grower, then the solution is to change varieties, crops, or growing methods that are more suitable for that region.

Advection freezes are characterized by the lack of inversion and without an inversion layer most methods of frost protection become ineffective. Those that are effective are over tree irrigation which is also extremely dangerous to use under these conditions, and wind machines or helicopters and heaters used together which may not be cost effective if needed for prolonged periods of time.

When using over tree or over vine irrigation and the sprinkler system fails for any reason during the event such as pump failure, running out of water or frozen/broken sprinkler emitters, catastrophic losses will occur. During advection freeze events the temperature may not rise after the night like during a radiation frost. It is not unusual for the temperature to stay below freezing during the day and the night sometimes for several days or weeks at a time. The danger in using over tree water for protection under these conditions is three fold. First, this will require significant amounts of water as the sprinklers might need to be run virtually 24 hours a day, maybe for several days at a time. This will increase the likelihood of simply running out of water leading to catastrophic consequences.

Second, protection with over tree water is dependent on the heat released from water drops freezing on the ice that encapsulates the buds in order to keep the ice at a safe temperature. The water volume from the sprinklers is constant, and the ice that has formed has no chance to melt, so the size of the ice mass continues to increase on the buds. Due to the ever increasing mass of the ice, it takes more heat, and therefore more water, to keep the ice at a safe temperature. At some point, there will not be enough water sprinkled on the ice in order to keep it at a safe temperature and the ice will start to dry and evaporate. This will cause the ice to lose heat and become much colder than the surrounding air, causing catastrophic losses.

These losses can be in excess of any loss that might have happened if no protection at all was used.

Thirdly, because an advection freeze can be accompanied by winter storm conditions, another problem caused when using over tree irrigation is windy conditions. Wind blowing on over tree water will cause evaporation to occur which can further cool the air temperature and the ice on the buds causing greater damage than what might have occurred if no protection at all was used.

Another method of protection that may be useful during advection freezes is heat and wind machines or helicopters working together. In this method, the heat created by burning fossil fuels is forced back down into the orchard by high velocity wind machines. Because under advection conditions there is no inversion, the air gets colder higher from the ground. This permits any heat generated near the ground to escape quickly into the atmosphere providing no benefit in the orchard and wind machines or helicopters used alone will blow air that is colder than what is already in the orchard downward. By using heaters placed strategically around the wind machine, the escaping heat will be pushed temporarily back down into the orchard to mix with the cold air.

The downside and main risk to this method is the high cost. Neither wind machines nor heaters can be effectively used alone in advection conditions, they must be used together. The high acquisition costs of wind machines and helicopters, and the high operating costs of heaters in conjunction with the indeterminate length of advection freezes may not prove to be cost effective. It will require over 40 gallons of fossil fuel per acre each hour to protect with this method, plus the cost of wind machines or helicopters.

Because of the these risks, in the event of an advection freeze the grower should perform a cost-benefit analysis and then make an informed decision on whether or not to attempt to frost protect or simply ride out the storm and hope for the best.

The next danger point of an advection freeze comes as the storm passes. Behind some winter storms is a cold, dry, clear mass of air. As the storm passes, even if there is no frost damage due to advection freeze, a radiation frost event can move in.

Radiation frost – Micro climate frost events

Radiation frost events are site specific and are caused by cold air accumulation. Other than a consistent correlation between temperatures in micro climates in the same regional climate, there is no connection between micro climate areas.

Under radiation conditions, whether it is in the winter, spring, or fall, the main cause of frost damage is cold air accumulation. The differences in temperature between the damaged areas and the non-damaged areas are about the same during the winter as they are under the same clear sky, no wind conditions as during the spring or fall. During the winter time, the temperature on the hill where there is no frost damage for example might be 18 deg. F, while the temperature in the frost pocket is 14 deg. F, 4 degrees less. During a spring frost event the temperature on the same hill might be 32 deg. F and in the same frost pocket it is 28 deg. F, still 4 degrees less. Of course, during the winter time the plants are dormant and can take much lower temperatures than in the spring or fall.

The 'kill' temperature for a plant during the wintertime when it is dormant will vary throughout the season. The critical temperature will vary up or down depending on the preceding 7 to 10 days weather. If there is an unusual warm spell for several days, the plants will begin to come out of dormancy and the plants can be damaged by significantly higher temperatures than if the preceding weather was normal or very cold. It is usually not the coldest night of the winter that causes frost damage, but the night where the temperature goes back to season normalcy, after several days of unseasonable warm weather.

Radiation frosts are characterized by a low dew point, clear skies and no wind conditions. Under these conditions the ground will lose heat through a process of long wave radiation. The ground will then cool. The cool ground will then cool the atmosphere from the ground up. The colder air molecules are heavier and denser than the warmer ones, and so will stay closer to the ground. This will cause the air to stratify into layers separated by temperature and density. Because colder air is heavier and denser than warmer air, the coldest air is closer to the ground and the temperature will rise as the distance from the ground increases. During these periods of stratification the air layers become a stratified fluid.

The heavier, colder air layer along the ground is affected by gravity and so will begin to move downhill if there is a slope. This downhill movement is called katabatic or down slope wind. It is this process that leads to the accumulation of cold air that is the main cause of frost damage. As the coldest air moves down hill and merges with other cold air streams, accumulation of the cold air will begin to take place. It is this accumulation of cold air that is directly responsible for radiation frost damage. Several factors must be present at the same time in order for a radiation frost to happen.

Factors that Need to be Present to have a Radiation Frost Event

There are several factors that need to come together at the same time for a damaging frost event to happen. The odds are against all of these factors coming together at the same time making a frost event unlikely on any given day. Areas that accumulate cold air due to natural inflow coupled with insufficient drainage are more likely to have a frost event than areas that do not normally accumulate cold air. Accumulation areas receive and hold cold air that is generated during the frost night. There are 5 major factors that need to be present to cool the ground sufficiently in order to generate an adequate amount of cold air to cause a damaging frost event.

1. Clear skies- Clear skies allow heat from the ground to escape into the upper atmosphere. As heat is lost from the ground through long wave radiation, and there is no cloud or fog obstruction, the ground will continue to lose heat and cool. Clouds or fog will reflect or absorb these radiation waves slowing the heat loss from the ground and keeping the ground warmer. It is the ground that cools the lower layer of the atmosphere resulting in the stratification of the atmosphere. Without stratification, there is no difference in the density of the air molecules and so there is no gravity influence on the heavier molecules along the ground. Without gravity influence and the resulting katabatic flow of cold air into areas with insufficient drainage, there can be no micro climates.

2. No wind. Windy conditions will mix the atmosphere retarding heat loss into the upper atmosphere. The mixing of the atmosphere prevents separation of the air layers and stratification. Without stratification, there is no difference in the density of the air molecules and so there is no gravity influence on the heavier molecules along the ground. Without gravity influence and the resulting katabatic flow of cold air into areas with insufficient drainage, there can be no micro climates. Most frost prone areas are in wind protected places.

3. Low dew point- The dew point is the temperature where the relative humidity reaches 100%. This is the saturation point of air where it reaches its maximum potential to hold water vapor. At this point condensation will take place causing dew droplets to precipitate from the air. The absolute volume of water vapor that can be held by a specific mass of air is dependent on the air temperature. The warmer the air temperature, the more mass of water vapor can be held. As the air cools, its ability to hold water vapor decreases while the absolute volume of the water vapor remains the same. The relationship between the absolute mass of water vapor (absolute humidity) and the maximum potential of the air to hold vapor at a given temperature is the relative humidity (RH). At 100% RH, condensation or dew begins to occur. This condensation process of turning vapor into liquid releases significant amounts of heat that is stored in the water vapor back into the air. It is this release of heat that will cause the air temperature to generally stabilize at or near the dew point. A dew point near or below the critical temperature of the protected crop is necessary in order for the air temperature to fall low enough to cause frost damage.

4. Soil Temperature- It is the ground that cools the air that causes frost damage. The soil temperature at the surface must be low enough at sunset to get cold enough that it will be able to cool a sufficient mass of air along the ground to cause damage. When there are clear skies and no wind, the ground will lose heat and begin to cool. The ground loses heat at a specific rate depending on factors such as the density of the soil and moisture content. The warmer the soil is at the start of the night, the warmer it will be when the sun comes up. The ground cools the air, so even if there are clear skies, no wind and a low dew point (at or near the lethal temperature for the crop that needs to be protected), frost damage will not occur if the soil temperature at the surface is high enough at sunset to avoid getting cold enough to cool a sufficient amount of air to submerge plant tissue in lethal cold air before the sun comes up. The actual soil

temperature required to cause damage will vary depending on soil type, cover crop, ground maintenance and moisture content. A dark, hard soil will lose heat more slowly than a light sandy or disked soil, so even if all of the conditions are the same in two different areas of an orchard, an area with sandy soil may experience frost damage while the rest of the orchard will not.

5.   Air temperature- The air temperature at sunset must be low enough to cool sufficiently to lethal temperatures. Even with clear skies, no wind, low dew point and cool ground temperatures, if the air temperature is high enough when the sun goes down, the above factors may not combine to lower the air temperature enough to cause frost damage.

An especially dangerous time for radiation frost comes at the end of a cold winter storm. Even if freezing temperatures are not reached during the storm, behind such storms is often a mass of clear, dry air (low humidity). As the storm moves out, the sky becomes clear and windless.

Winter storms are associated with clouds and wind that will prevent the ground from absorbing heat. The surface ground temperature cools because there has not been sunlight to warm it, possibly for several days. And due to the storm, the air temperature is also cool.

When a storm clears during the night and the wind stops, radiation losses begin immediately. The ground temperature, air temperature and humidity are all starting at low points and the resulting drop in air temperature can be very quick and very dramatic.

In the following days, after clear, sunny weather there is less risk of frost even if all of the other conditions are the same. This can be attributed to the ground having time to absorb heat, and the air temperature increasing. With each passing clear day, the risk of frost becomes less.

Differing soil types and densities will lose heat at different rates. Heavier, denser soils have better the heat retention than lighter, sandier type soils and less risk of frost. Under the same conditions, a heavy clay soil may not lose enough heat to pose a frost risk where a sandy soil will. All areas of the orchard will lose heat at the same rate after compensating for differences in soil density, moisture content and cover crops.

There are other negative effects on crops in micro climate cold air accumulation areas besides just frost damage. Accumulation areas are colder and will precipitate more water due to condensation or dew than warmer areas. This can help to cause cherry cracking, mold and mildew problems, root fungus and other diseases related to excessive moisture.

In many instances, the colder accumulation areas will experience a later bud break and possibly even later harvest dates. This is because the plants growing there will acclimate to their colder micro climate. This effect may cause problems with spraying, fertilizing, and harvest schedules.

Chapter 3:   Building Blocks of Frost

Inversion layer

An Inversion layer is created when conditions permit a deviation from normal atmospheric cooling related to a rise in elevation. In normal atmospheric conditions the temperature decreases 3.5 deg. F per 1000 ft in elevation. In inversion conditions, the temperature increases as the elevation from the ground increases. [5]

During periods of inversion, a warmer stratus of air exists over a cooler mass of air. Nighttime (nocturnal) inversions are caused by radiation heat loss from the air and ground, while daytime (diurnal) inversions are caused by lower layers of air being cooled by a large body of cool water or cool ground and then moving horizontally displacing warmer air layers. Daytime inversions are more common along coastal areas. Heat loss from the air and ground during a radiation event cools the ground and the lowest level of the atmosphere, known as the 'laminar boundary', which in turn cools the atmosphere from the ground up.[6]

The inversion layer is divided into two parts. The lower inversion, and the upper inversion. The upper inversion layer exists at a height of about 300 ft to 900 ft. above the regional floor. This is a large, very stable mass of air that maintains the highest overall regional temperature throughout the night largely because it is not affected by the ground cooling.   There is not much temperature difference throughout the upper inversion layer. At an elevation about 900 - 1000 ft above the regional floor, the temperature will start to drop. This cutoff point is a thermal boundary and is considered the 'ceiling' of the inversion layer. Above this layer of the atmosphere is affected first and most because it is the closest to the cold ground. Above the ceiling, the atmosphere turns back to advection and the normal rules of temperature loss in relation to altitude apply.

The lower inversion layer is from the ground to 300ft. When the sun goes down, the temperature in the lower inversion is roughly equal to the temperature in the upper inversion. The temperature in lower inversion layer will drop in relation to the regional temperature throughout the night. The closer to the ground, the faster and greater the temperature will drop because it is the ground cooling that causes the air to cool. The temperature at 300 ft will not drop below regional temperature much if any during the night, where the temperature at 5 ft from the ground can drop to below as much as 18 deg. F from the regional temperature before the sun comes up. In areas of cold air accumulation, the temperature can be as much as

---

[5] Vineyard Site Selection

Tony K. Wolf and John D. Boyer, Professor of Viticulture and Lecturer, Virginia Tech

[6] **Boundary layer climates** (Second edition). By. T. R. Oke. Methuen. 1987

another 12 deg. F lower than the lowest regional temperature. It is not unusual to have a 30 deg. F difference in temperature under radiation conditions within a regional climate! When the sun comes up and the sun's rays hit the ground, radiation losses from the ground are immediately stopped. The sun will then begin to heat the ground with short wave radiation which will quickly destroy the air stratification and raise the temperature.

Example temperature distribution throughout a regional

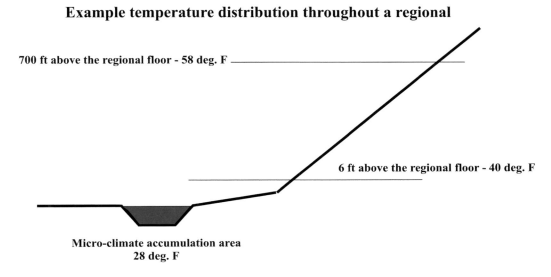

700 ft above the regional floor - 58 deg. F

6 ft above the regional floor - 40 deg. F

Micro-climate accumulation area
28 deg. F

Fig 1 - Example distribution of temperature throughout a regional climate

In the following graph, three data loggers placed in an orchard show the relationship under radiation conditions of different areas. Data loggers A and B are placed in separate well drained areas, while data logger O is placed in an accumulation area. As the accumulation area fills with cold air, there is a consistent temperature differential between the areas.

When the sun rises in the morning, stratification is destroyed and the temperature rises quickly.

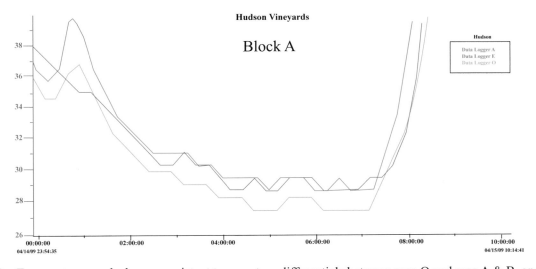

Fig. 2 – Temperature graph shows consistent temperature differentials between area O and area A & B, and quickly rising temperatures when the sun rises.

Thermal Gradient

A thermal gradient is the difference in temperature at a specific vertical distance and defines the strength of the inversion and atmospheric stratification. For example a 1 deg F rise in temperature from 3ft to 6 ft is a thermal gradient of 1 deg. F/3ft. The thermal gradient will tend to spread out as the distance from the ground increases and become less pronounced. The second 3ft rise in elevation will have less temperature increase and each subsequent 3ft rise will have less temperature rise than the previous one. Inversion strengths can be classified by comparing the temperature at 3ft above the ground to the temperature at about 40ft above the ground;

Strong inversion - 8 - 10deg F difference
Medium strong - 6 - 8 deg F difference
Medium   - 4 - 6 deg F difference
Weak   - 2 - 4 deg F difference
No inversion   - 0 - 2 deg F difference

Most frost systems work best with a stronger inversion.

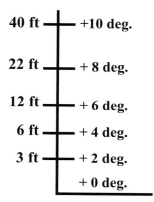

Fig.2 - Typical thermal gradient in a stratified atmosphere. As the distance from the ground increases, there is a larger gap to achieve a 2 deg. temperature increase. The temperature differences show atmospheric stratification.

Atmospheric stratification and Thermal Boundary lines

Stratified atmosphere is the layering of the atmosphere due to differences in density and temperature. Density is inverse to temperature- the lower the temperature of the air molecules, the heavier and denser they become. Density is not the same as viscosity, the colder the air molecules, the denser but less viscous they become. Since density is related to temperature, the layers can be identified and separated by temperature differences, as in Fig 2. It is not possible, during radiation conditions, for a warmer molecule to be below a colder molecule under natural conditions. The colder air will always be closer to the ground unless modified.

Visual image: On a radiation night, clear skies, no wind, go to a relatively flat area in an accumulation basin. Hold a thermometer at 3ft from the ground and record the temperature. Then raise the thermometer at 1ft increments and see the temperature differences.

Thermal boundary lines are separation points in the atmosphere due to the differences in the density and temperature of the layered air molecules. The differences in the characteristics of the air layers can cause differing reactions on either side of a thermal boundary line to forces such as gravity or cold air accumulation.

A thermal boundary is sometimes referred to a as a 'ceiling', and can be used in an absolute or relative context, such as the natural absolute ceiling at the top of the upper inversion layer, or in a relative context such as the artificial ceiling created by heaters where warmed molecules stack. At the absolute ceiling at the top of the upper inversion, the temperature above the thermal boundary begins to drop and the temperature below the thermal boundary also drops. In this case the thermal boundary is the warmest point; both sides of the line get colder as the distance from the thermal boundary increases. In the case of an artificial ceiling where heated air molecules will stack, the molecules directly below are the same (or possibly even a slightly higher) temperature as the thermal boundary line. Before the warming with artificial heat takes place the temperature is colder below this point and warmer above- hence no thermal boundary. This line of demarcation marks the separation point where the temperature rises on both sides of the lines due to the introduction of artificial heat sources.

The term 'thermal boundary layer' also refers to an absolute thermal boundary where there is a physical separation such as the point of gravity influence (the point where molecules below are affected by gravity and moving, and those that are above and static), or in a relative context such as the separation of lethal and non – lethal temperature air mass, even though both are moving or non-moving.

Under radiation conditions, the atmosphere will separate or stratify into layers defined by temperature and density unless some other phenomenon destroys the stratification. Clouds and wind will destroy stratification, as will surface fog and overhanging trees. This is the result of the radiation waves being absorbed and reflected back into the ground slowing the ground cooling and inhibiting heat loss into the atmosphere. Crops that are planted under an overhanging tree will often not suffer frost damage for this reason. Crops that are planted in areas where the thermal boundary layer can flow more freely and flatten out, such as the first row or two along a wide road will also tend to avoid frost damage as cold air is free to drain off.

A thermal boundary layer can also be a demarcation line where there is a difference in dynamics of the molecules on either side of the line caused by a difference in temperature and density of the separated molecules. In fig. 3, the thermal boundary line is defined by the separation of the higher, warmer molecules that are too far from the ground and not dense enough to be affected by gravity, and the colder, heavier molecules in the lower stratus that are affected by gravity. The molecules above the thermal boundary line (disregarding for turbulence created by the down slope air movement) that are not affected by gravitational pull will remain static until they cool and fall below the TBL. All the molecules below the TBL are affected by gravity and will flow downhill. The colder, denser molecules closest to the ground are more affected and will move faster than the lighter, warmer molecules.

Thermal boundary lines can be vertical or horizontal and can exist in a static mass of air such as a build up of cold air over the height of a barrier, or a thermal boundary can exist in a dynamic flow as the separation point between the static and dynamic molecules.

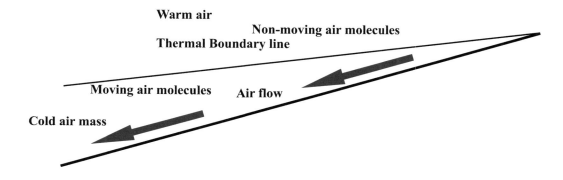

Fig. 3- Absolute thermal boundary line in a dynamic flow showing separation of the moving and non-moving air molecules.

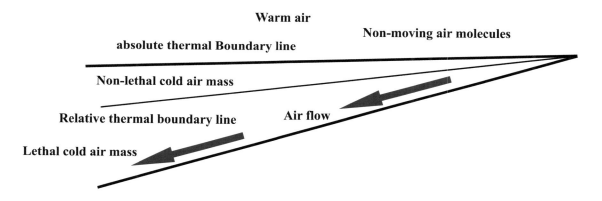

Fig 4 – Relative thermal boundary line is the point of lethal and non lethal air mass while the absolute thermal boundary line is the separation point of dynamic and static air molecules.

The height of the thermal boundary line in a dynamic air flow will vary depending on the angle of slope, total mass of cold air entering the flowing zone from outside the flow, ground conditions, cover crops, and temperature of the cold air along the ground.

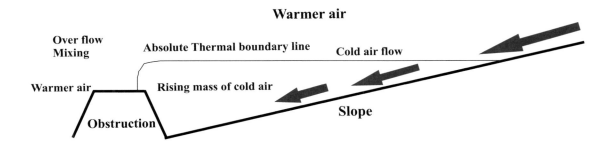

Fig. 5 - Thermal boundary line in a static air mass. A vertical boundary exists in a relationship to the height of the obstruction and the angle of slope approach.

In fig. 5 cold air flowing into a basin that is blocked by an obstruction can build up a thermal boundary line that is vertical and higher than the obstruction itself. The actual height of the thermal boundary layer is determined by the angle of approach to the obstruction and the height of the obstruction. A lower slope angle of approach will raise the thermal boundary layer higher in relation to the height of the obstruction than a steep angle of approach.

As the colder air enters the basin area, it accumulates under the existing air molecules and pushes them upward, sometimes 2 to 3 times the height of the barrier. The thermal boundary layer in this case is the height at which the cold air entering the basin displaces the air that existed in the basin.

Down slope (Katabatic) wind[7]

Katabatic air flow is downhill air movement caused by gravity.

In a stratified atmosphere, the colder, heavier air molecules close to the ground are affected by gravity. It can take very little slope angle for cold air molecules along the ground to begin to flow downhill. The actual slope angle required to move air is dependent on factors such as the smoothness of the ground, cover crops and air temperature at the ground. It may take less than ½% slope angle to move air along the ground. The smoother the surface, the less friction will

[7] **Dynamical Processes in Undisturbed Katabatic Flows**
  Gregory S. Poulos, Thomas B Mckee, James E. Bossert, and Roger A Pielke
  Los Alamos National Laboratory, Los Alamos, N.M., Colorado State University, fort Collins, CO.

Thermal Belt and Cold Air Drainage on the Mountain Slope and Cold air Lake in the Basin at Quiet , Clear Night
Yoshino,.M.M., Prof. Dr. Institute of of Geoscience, University of Tsukuba, Tsukuba, Ibaraki, 305 Japan

need to be overcome, encouraging air flow. Cover crops, row berms, and other obstructions that run perpendicular to the slope will impede the flow of air along the ground causing a buildup of cold air and a higher thermal boundary line.

Katabatic air flows on an uncovered or slightly treed surface @ 10% or more slope angle can reach speeds of 6 mph. The lower the angle of the slope and/or the heavier the vegetation on the slope, the slower the air will flow.

Under perfect uninterrupted radiation events, cold air from long distances can reach the damage area. A wind or clouds interruption will mitigate the damage by dispersing the advancing cold air and the process must begin again.

Air will flow faster on hard, smooth surfaces such as paved roads than on irregular, soft ground. Keeping the surface below a growing area groomed to provide maximum flow of cold air out of the growing area may help decrease the possibility of frost damage. Likewise, installing obstructions uphill from the growing area may also decrease the risk of frost damage. Obstructions and air flow enhancements should only be done after a site analysis on the area is completed.

The air temperature is related to the density and viscosity or thickness of the air. The colder the temperature, the denser the air becomes while becoming less viscous. Colder air is heavier, denser, and more affected by gravity.

The natural drainage of cold air that occurs on hillsides is the result of katabatic flow. Steep hillsides are regarded as frost free not because the ground on the hill loses less heat than the ground in the low accumulation area. It doesn't- the ground in all areas of the orchard loses the same amount of heat. The reason that hillsides are naturally frost free is because the cold air generated there flows off sufficiently fast to avoid a deep thermal boundary line. By keeping the thermal boundary line and thus the lethal cold air mass lower than the susceptible plant tissue, there is no frost damage. However, if these cold air flows combine into converging currents, or are obstructed in other ways, then accumulation of cold air can develop.

Visual image – To see a Katabatic air flow and a thermal boundary line on a radiation night – clear skies, no wind - go to a relatively steep slope that is bare or only lightly covered with vegetation and go a partial way up the hill. Light a candle and hold the flame close to the ground. You will see the flame bend down hill. This is caused by the down slope or katabatic air movement. As you raise the candle up away from the ground, the flame will get straighter and straighter until at some point a few feet or several feet from the ground, the flame will stand perfectly straight. This is the separation point of the air molecules that are affected by gravity and flow down hill, and those that are too light and too far from the ground to be affected. This demarcation point is a Thermal Boundary Line- all the molecules below this point are affected by gravity and all those above this point are not affected.

Since katabatic air flows are driven by gravity, which is probably the most consistent force on earth, these flows can be tracked, anticipated and manipulated. By using 3D modeling and airflow vector models, cold air flows and accumulation can be calculated and modified by using specially designed barriers and diversions and through Shur Farms Cold air Drain® technology.

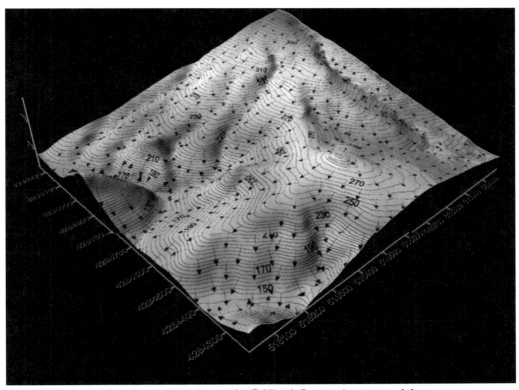

Fig 6- Shur Farms Frost Protection® 3D Airflows w/vector model

Return Period of a Frost Event

The timing and intensity of any given frost event is random, however, the frequency or 'return period' of these events when measured over a long period of time is subject to the laws of statistical probability. This is known as a stochastic event.

Just like any stochastic event, a damaging frost episode will happen with a certain frequency in a particular area the same way the frequency of other stochastic events like earthquakes and floods will happen. Because of this, the Return Period (RP) – or likelihood of a frost event in any given year can be measured and translated into a risk factor. Even though it is not possible to determine if there will be a damaging frost in any given year, the statistical relevance is apparent in a 10 year or longer cycle.

The frequency or probability of a frost event happening at a specific time is measured within blocks of time during the year, commonly a 2 week period. This period is then compared with an analysis of the previous corresponding periods in a 10 year or longer cycle. For instance the time period of April 1st -15th can be compared to the previous 10 years during April 1st – 15th period. If there has been a frost event in 3 out of the last 10 years during April 1st -15th, then there is a 30% chance that there will be a frost in the coming period. A comparison is then made for the period April 16th -30th, then May 1st -15th and so on throughout the entire frost risk period. Any number of frost days during the analyzed period still counts as one. Even if there were 6 frost nights in that period during one year, and 3 in another during those 3 frost years, there is still a 30% chance of frost in the coming year because 7 out of 10 years were frost free.

Micro climates are the result of cold air build up due to insufficient drainage. Any factor that increases or decreases inflow of cold air into the area, or outflow of cold air from the area will change the micro climate. These factors can include the addition or removal of vegetation, buildings or other structures, roads or downstream and upstream obstructions.

If the orchard or vineyard has been in existence for a long period of time and the topography, vegetation, soil type and other factors inside and outside the area (to the extent that it affects the orchard) has not changed, then it is possible to compare the frost history of the orchard and make a statistical analysis on the future probability of a frost event. If relevant factors have changed, this will affect the results because the micro climate in the orchard might have changed.

In the following charts (Fig. 7), B Block - Lower Swale experienced frost events in the period March 15th -30th, seven times in the last 10 years. The same area experienced frost events in the period April 1st -15th, three times in the last 10 years and once in the period April 15th -30th. After May 1, no frost events were recorded.

We can make an analysis that in B block, Lower Swale there is a 70% chance of a frost event during March 15th -30th, a 30% chance of a frost event April 1st -15th, a 10% chance April 15th -30th, and little or no risk after May 1.

In L Block, below Dan's House there is a 40% chance of a frost event during March 15th -30th period, a 30% chance April 1st – 15th period and little or no risk after April 15th.

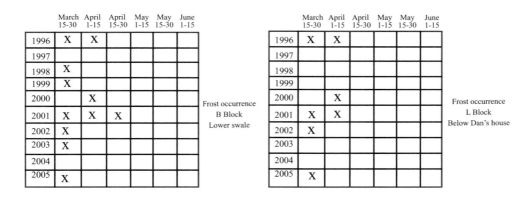

	March 15-30	April 1-15	April 15-30	May 1-15	May 15-30	June 1-15
1996	X	X				
1997						
1998	X					
1999	X					
2000		X				
2001	X	X	X			
2002	X					
2003	X					
2004						
2005	X					

Frost occurrence B Block Lower swale

	March 15-30	April 1-15	April 15-30	May 1-15	May 15-30	June 1-15
1996	X	X				
1997						
1998						
1999						
2000		X				
2001	X	X				
2002	X					
2003						
2004						
2005	X					

Frost occurrence L Block Below Dan's house

Fig. 7- Frost occurrence chart

There is a relatively consistent temperature difference during a radiation event (clear skies, no wind) between micro climates within the same regional climate. Areas that have been modified or newly planted and have no relevant frost history can still be analyzed.

An air flow model such as in Fig. 6 is created which will identify the areas of potential cold air accumulation and dynamic air flows. Data loggers are then strategically placed in the orchard to determine temperature differences within the orchard. By comparing the temperature differences between different areas in the orchard, a micro climate map can be made of the entire orchard which will show the coldest and warmest areas in relation to each other. By correlating those temperatures with a known source of historic temperature data such as a CIMIS station or ADCON station, conclusions as to frost risk can be made with a full frost risk analysis.

Another method is to compare the temperatures of the new site to a site close by (in the same regional climate) that has a known frost history for the same crop and variety. By using a known area as the control and comparing temperature differences between that area and the areas in the new orchard, a good indication of the viability of growing the same crop can be made.

Because the probability of any magnitude of frost event (or no frost event) in a specific year is the same as any other year, including a 100 year or greater frost event, the success or failure of a frost protection system might not be able to be determined on a year to year basis.

Frost Damage vs. The Potential for Frost Damage and Super Cooling

Simply put, frost damage occurs in plants when water in the cells of the plant crystallizes into ice and ruptures the tissue. Until this happens, the plant will remain undamaged.

The 'critical' or 'lethal temperature' is the point at which crystallization of ice inside the plant tissue can occur. This point is variable depending on the species, variety, phenology and other factors of the plant including general health and size. When a plant reaches or falls below its critical temperature it now has the 'potential' for frost damage.

It is often said that the freezing point of water is 32 deg. F, but this is incorrect. 32 deg. F is the melting point of ice. Water will sometimes not freeze until well below 32 deg. F, but it pretty much will always melt at temperatures above 32 deg. F.

The absolute freezing point of water is (minus)-40 deg. F or Celsius, at this point Fahrenheit and Celsius are the same. This temperature is called the homogeneous freezing point and water molecules will freeze without any nucleus or outside stimulus. In order for water to freeze at a higher temperature, then there must be something to act as a nucleus for crystallization. Crystallization around a nucleus is called heterogeneous ice formation and the most common nucleus is a bacterium that resembles an ice crystal referred to as an ice nucleating bacteria (Pseudomonas Syringae). [8]

Super cooling is the ability of a plant or other media to go below the critical temperature required to induce heterogeneous ice crystallization, and not freeze.

Visual image - Take a glass of de-ionized water and place it in a freezer. At about 25 deg. F, remove the glass and carefully place it on a table- the water will not be frozen. If left undisturbed, the water temperature will slowly rise to above the freezing point and then it will not be able to freeze. However if the water is agitated by shaking or stirring, super cooling is stopped and ice crystallization will start immediately!

In the above visual image, the water at 25 deg. F has the potential to freeze, but some of the factors that cause ice to crystallize are not yet present, However movement of tissues that contain super cooled water can cause ice nucleation, freezing and death.[9]

The longer that a crop remains below the critical temperature zone, the greater the risk that ice crystallization will start. There may be more danger for a crop to be at 29 deg. F for a four hour period than to be at 28 deg. F for 30 minutes.

A concept of frost protection is to lessen the time that a crop is exposed to critical temperatures, and by doing so, lessening the risk of frost damage.

Once ice crystallization starts, there seems to be a chain reaction and rarely will only one plant be affected. Usually a large area is affected and sometimes the spread of crystallization (and therefore frost damage) will stop due to a break in the field such as a row divide or road.

---

[8] **The use of genetically engineered bacteria to control frost on strawberries and potatoes. Whatever happened to all of that research?**
R.M. Skirvin, E. Kohler, H. Steiner, D. Ayers, A Laughnan, M.A Norton, M. Warmund   Univ. of Illinois, Dept of Natural Resources and Enviornmental sciences, Univ. of Missouri, Dept of Horticulture

[9] Practical Considerations for Reducing Frost Damage in Vineyards   M.C.T. Trought, G.S. Howell and N. Cherry Report to New Zealand Winegrowers: 1999

Chapter 4: Myths of Frost Protection

1. Myth - Cold air that causes damage 'falls' from the sky and settles in the low spots.

Reality- Cold air that causes frost damage is cooled by the ground and flows along the ground downhill into accumulation areas due to gravity. The cold air is the result of the ground losing heat through long wave radiation during nights that have no wind and no clouds. As the ground loses heat and cools, it in turn cools the air closest to the ground. Colder air molecules are heavier and denser than warmer ones and will stay closer to the ground. The lower, colder molecules are affected by gravity. Cold air will begin to move downhill until insufficient drainage conditions such as physical obstructions or low slope impedes the flow and causes the cold air to accumulate. When the cold air builds up deep enough to submerge plant tissue frost damage may occur.

Cold air masses that come in over the ground due to a weather system and are below the critical temperature for the crop to be protected are rare. Even though there are the characteristics of radiation frost- clear skies and no wind, this is a regional frost event that is the result of a cold mass of air associated with a winter storm or advection freeze.

When these types of events happen more frequently than is economically viable, then the grower must consider that the crop he is trying to grow is not suitable for the region and make adjustments.

2. Myth – Blowing wind over plants will prevent frost.[10]

Reality- Blowing air or otherwise disturbing plant tissue while it is in a super cooled state will promote ice crystallization in the cells by stopping the super cooling process and promoting ice crystallization in the cells.

In order for wind to be effective, there must be warmer air from the higher layers pushed down and mixed with the colder air to increase the overall air temperature, or heat must be added to the wind blowing. It is the rise in temperature produced by mixing warmer air molecules with the colder air molecules close to the ground that protects against frost, not the movement of air.

---

[10] Practical Considerations for Reducing Frost Damage in Vineyards   M.C.T. Trought, G.S. Howell and N. Cherry
Report to New Zealand Winegrowers: 1999

3. Myth – Smoke protects against frost[11]

Reality- Smoke does nothing to stop or slow heat loss from the ground. On the contrary, smoke does prevent heat from reaching the ground when the sun comes up in the morning. Smoke is not only useless in protecting against frost, it is a detriment.

The ground loses heat through long wave radiation which is 10 -15 microns in diameter. Smoke particles are > 1 micron in diameter. Smoke does nothing to stop, absorb, or reflect radiation losses from the soil. It is invisible to the long wave radiation and has no impact on ground cooling or the resulting cooling and stratification.

When the sun comes up in the morning, the heat radiated back to the ground from the sun is short wave radiation. Short wave radiation is approximately 1 micron in diameter and is effectively blocked by smoke. This effect will delay the warming process.

Smoke can be useful when using heaters for frost protection to indicate the height of the stacking of the molecules. When using heaters for frost protection, the stacking point of the warmed molecules is a ceiling. When heaters warm the air, the difference in temperature between the heated air molecules and the natural surrounding molecules will cause the warmer molecules to rise. This effect is called 'heat buoyancy'. The warmer air molecules will rise, and quickly lose heat, to a height where they are 'neutrally buoyant', or the same temperature as the air around them. For example, since a molecule of 32 deg. F will not push away another molecule of 32 deg. f (the same temperature and density), the upwardly rising molecule will stop when it is blocked by a natural molecule of the same density- a process called 'stacking'. This point at which the heated molecules stop rising is the 'ceiling'. The process of heating molecules and stacking them at a given height creates a thermal boundary line. The newly stacked molecules will then begin to cool and move downward again in a convection manner until they are reheated by the heat source and rise again. Smoke particles will rise with the rising warm air molecules and will move horizontally along the thermal boundary line giving a visual representation of the stack height of the heated molecules. The temperature from the heaters can be adjusted up or down as necessary to achieve the proper ceiling height.

---

[11] **FROST/FREEZE PROTECTION FOR HORTICULTURAL CROPS**

Katharine B. Perry, Ph.D.
Extension Agricultural Meteorologist College of Agriculture & Life Sciences
North Carolina State University

Visual image - One use for smoke particles is that they will move horizontally along the demarcation line (thermal boundary line) and give a visual representation of height of the 'ceiling'.

4.   Myth – Frost on the ground indicates a frost problem.

Reality- Ice on the ground is not an indication of frost risk. It is the deepening mass of lethal cold air that causes frost damage, not the temperature of the ground.

The ground in all areas of the orchard or vineyard will give up heat at the same rate (providing all other factors of soil density, moisture content, etc. are the same), and the ground will cool at the same rate. Areas that are well drained are not generally at risk for frost damage because there is no cold air accumulation. The ground temperature on a hillside is approximately equal to the ground temperature in the lower bowls, but the cold air generated by the cool ground on a hillside flows off and does not build up deep enough to submerge plant tissues. The natural drainage capacity on the slopes prevents cold air accumulation and avoids the buildup of cold air.

5.   Myth – Removing the downhill barriers such as trees and brush will improve cold air drainage.

Reality- Downhill barriers can sometimes block cold air that builds up below from flooding back into the orchard as well as block the cold air generated from above from draining out.

Cold air accumulates in areas of insufficient drainage. If the lower basin area below an orchard has cold air from surrounding sources that exceeds the capacity of the basin to drain itself, cold air will build up from below. This rise in the height of the cold air mass in the lower basin will cause cold air to flood back into the orchard unless it is blocked from below and removed. A barrier along the lower portion of the orchard may help to inhibit cold air entrance into the vineyard and actually be a benefit.

One way to tell when there is insufficient drainage and flooding from below is that the air temperature in the lower basin will be equal to or lower than the temperature in the orchard, and the temperature in the lower basin will drop prior to the temperature in the upper orchard.

In other instances, the basin below the orchard is well drained and the cold air that is causing frost damage is originating from the orchard area itself and/or areas above the orchard. Cold air flowing downhill may not be building deep enough to cause damage until the downhill barrier obstructs the natural drainage causing accumulation of cold air in the orchard, a cold air lake.

If there is adequate drainage below and cold air accumulation is caused by the barriers, the temperature in the lower basin will be higher than the temperature in the orchard.

Often, both cases will exist. Under mild frost conditions there may be adequate drainage in the lower basin due to a smaller volume of cold air being generated from the surrounding areas that flows in there, while under severe frost conditions the lower basin will be overwhelmed and accumulate cold air due to the larger volume of cold air flowing into it. When the inflow of cold air exceeds the capacity of the basin to drain itself, accumulation will occur.

Where there is insufficient drainage below, removing barriers will provide no benefit and could make the situation worse.

2.   Myth – Cold air is like water

Reality – While there are similarities in the way that cold air flows under stratified conditions as compared with the flow of water, there are also some differences. Both water and cold air flows are caused by gravity and so they will both follow the path of least resistance. Some of the differences between air and water are;

1.   Water does not change viscosity with temperature change, until it freezes or evaporates.
2.   The density of air changes in direct relationship to its temperature. The colder the air, the more dense it is and the heavier the molecules. This characteristic guarantees that the coldest air molecules will always be lower than warmer molecules under natural conditions. The heaviest water molecules are at +4 C (approx 39 deg. F). As water cools on top of a lake, it becomes heavier and sinks. At lower temperatures, it becomes less dense and rises again. If not for this characteristic, ice would form at the bottom of pool of water, not on the top.
3.   The speed of air flowing downhill is determined by the slope angle, ground surface (smooth, slick surfaces produce less friction) and differences in viscosity caused by air temperature and absolute humidity. The speed of water flowing downhill is determined by slope angle and the ground surface. The temperature is not a factor.
4.   Cold air running into an obstruction can build up to 3 or 4 times the height of the obstruction before spilling over. The height that the cold air will build to is determined by the angle of the slope approaching the obstruction, viscosity of the air, and the volume of air blocked by the obstruction in relation to the available drainage capacity. The lower the angle of approach, the lower the air temperature, and the larger the mass of cold air blocked by the obstruction, the higher the cold air mass will build in relation to the height of the obstruction, and then further and higher back into the orchard. Water running up against an obstruction will build up to the height of the obstruction and then try to flow over. Water will build higher than an obstruction if the volume of the water blocked by the obstruction exceeds the available drainage capacity.

3.   Myth – More and hotter heat is better than less heat.

Reality- Applied heat that is too hot will rise quickly above the orchard and have no effect in the growing zone.

As cold air molecules near the heat source are warmed, the warmer molecules become lighter and less dense than the surrounding colder air molecules. This causes the warmer molecules to be thermally buoyant and to rise. The greater the difference in temperature between the heated molecules and the surrounding colder natural molecules, the more buoyant the warmer molecules will become. These molecules will quickly lose heat as they rise, and at some point above the ground, they will be the same temperature and density as the surrounding molecules.

The greater the difference in temperature between the warm molecules and the cold molecules the greater the tendency to rise, resulting in faster and higher ascendency of the warm molecules.

The height above the ground where the heated molecules no longer rise is referred to as the "ceiling" and this where the molecules "stack" under the ceiling. Since molecules of the same temperature will not push away each other, molecules of approximately the same temperature will stack under each other until the temperature in the entire area below the ceiling is equalized.

The ceiling height can be controlled by modifying the amount of heat or heat concentration coming from the heat source. A more indirect heat – or radiated heat – will impart the same level of heat energy (BTU's) over a larger area while turning the heaters down and burning less fuel will provide less BTU's. Either approach will provide less temperature difference between the heated molecules and the natural molecules resulting in less thermal buoyancy. This will result in a lower ceiling or 'stacking' height.

In most cases the ideal ceiling height is just slightly above the canopy.

Using more heaters, each with less fuel consumption will have the same effect as fewer heaters that have better radiation capability.

When the molecules at the ceiling begin to cool they start to fall until they are again heated by the heat source and the process repeats. This continuing circulation of molecules being re-heated, rising and falling is called "convection".

No real protection takes place until the convection process is complete.

If a molecule is super heated with a direct concentrated and non-radiating heat source such as a bonfire, then the heated molecules will rise to much higher level and may not have the time to cool and fall back down into orchard. In this case, the grower will only have succeeded in heating the atmosphere above the orchard, making no impact on the temperature inside the growing zone.

The limit of the temperature rise inside an orchard is the temperature at the ceiling. For instance, if the temperature at 15ft above the ground is 3 deg. F higher than at the ground, and this is where the ceiling is set, then the maximum rise in temperature inside the orchard will be 3 deg. F. In order to achieve a higher temperature rise, then the ceiling must be raised which would require more fuel and more time for the convection process.

The thermal gradient (the temperature rise at a given height increment) spreads out as the height from the ground increases. For instance there may be a 3 deg F. rise from the ground to 15ft, but the next 15 ft may raise only 1.5 deg. F, and the next 15ft after that may raise only .5 deg F. Fuel consumption rises exponentially to the rise in temperature required in the growing zone.

In all cases using heaters, control of the ceiling height to accomplish specific goals is essential and more heat is not always better.

4.   Myth – Heating water will improve the effect of under tree irrigation

Reality- Heating water prior to applying it for under tree irrigation, provides little or no benefit.

Under tree irrigation releases heat generated from the cooling and freezing of the water that is emitted or flooded over the ground. The heat that is released from the water is transferred into the surrounding air molecules, which then become thermally buoyant and rise until they are same temperature as the air around them. At this point they will stack under each other creating a ceiling and begin convection. The principle way that under tree water protects crops is the same as the principle of heaters. The main difference is the intensity of the protection. Molecules heated by gas burning heaters can rise high above the orchard canopy while molecules heated by water will tend to rise only a few feet from the ground. The temperature rise in the orchard in both methods is limited to the temperature at the ceiling height.

To accomplish heat release from the water in the most effective way, the water must convert to ice before it soaks into the ground. Heat released due to the water changing state from a liquid to a solid is called the latent heat of fusion. The following formulas illustrate the potentials:

Water state changes that will release heat and warm the air-
1 lb (approx. 1 pint) of water is cooled by 1 deg. F. = 1 BTU's heat released.
1 lb of water changes state from liquid to solid (ice forms) = 144 BTU's heat released.
1 lb of water is condensed from a gas state (as when dew forms) = 1077 BTU's heat released.

Water state changes that will use heat and cool the air –
1 lb (approx. 1 pint) of water is warmed by 1 deg. F. = 1 BTU's heat used.
1 lb of water changes state from solid to a liquid (ice melting) = 144 BTU's heat used.
1 lb of water is evaporated from a liquid to a gas (evaporation) = 1077 BTU's heat used.

Achieving a high conversion rate from liquid to solid, and condensation from water vapor to water are the most efficient ways of releasing the stored heat in water.

Heating the water will slow the time it takes for the water to reach freezing temperature. This delay will cause a lesser conversion rate to ice before the water soaks into the ground. When the water applied does not freeze, but instead soaks into the ground, the major source of heat and frost protection is lost.

5.   Myth- Heaters placed around a perimeter will cause heat to drift into the orchard

Reality – Heaters around a perimeter create a thermal barrier and blocks cold air trying to enter from outside.

A thermal "wall" is created by placing the heaters at a close distance to each other along a perimeter boundary, usually along the uphill side, or the side facing a prevailing cold air drift. The heat from the heaters will not flow downward into the orchard – heat only goes up, never down.

The effect is to heat the cold air moving in from outside the orchard (due to katabatic flow or from a prevailing drift), and block its entrance into the orchard by heating it and sending it upwards. Since it is this cold air that causes most or all of the damage inside the orchard, the avoidance of cold air entrance offers protection against frost damage.

The coldest air molecules are the ones closest to the ground. These cold air molecules will mostly avoid being heated and stopped because they will go under the heat wall, rendering this method only partly effective and hugely inefficient.

The same effect can be accomplished with forced cold air drainage and barriers at a fuel use ratio a fraction of that of heaters, and with much higher effectiveness.

6.   Myth – Natural ice formation on plants is good because ice formation releases heat and keeps the plant warmer.

Reality – Natural ice formation on plant tissues, or white frost, will cause damage at higher temperatures than black frost conditions related to low relative humidity.

During periods of low humidity where the temperature falls below the critical point but does not condense water or ice onto the plant, super cooling of the plant tissue is enhanced. The plant will experience less damage than if the dew point were higher allowing water and ice to form on plant. Surface water on the plant, such as is left on after a rain will act as a nucleator for ice within the bud. High relative humidity can have a similar impact on super cooling as does surface water. Ice crystal formation on the bud surface can serve to inoculate the tissue and canes.[12]

Natural formation of ice on plants due to water precipitation from the air is not the same as over tree irrigation.

_____

[12] Practical Considerations for Reducing Frost Damage in Vineyards   M.C.T. Trought, G.S. Howell and N. Cherry Report to New Zealand Winegrowers: 1999 (cited Johnson and Howell, 1981)

Chapter 5 – Frost Protection Tools

Within the categories of Passive and Active frost protection measures, there are three main sub-categories of frost protection – those methods that *protect against the cold air* by putting on a shield, those that focus on *preventing the area from getting cold* by warming the air, and those that focus on *making the plant less susceptible to the cold.*

Methods that *protect against the cold air* include over tree irrigation, trunk wraps, anti-frost insulation and foams. These methods are not reliant on an inversion layer and may have some effect under advection conditions. However, if advection freeze is a significant problem then the grower should be looking at planting crops or varieties more suitable to the region.

The second category focuses on *preventing the area from getting cold.* In this category there is wind mixing (wind machines and helicopters), heaters, under tree irrigation and cold air removal technology (Shur Farms Cold air Drain®). Passive measures that focus on keeping the area warmer include air barriers and diversions, soil management and cover crop management. These methods work by reducing the time that the crop is in a super cooled state or exposed to critical cold temperatures. Limiting exposure to lethal temperatures reduces the likelihood and statistical risk of actually having frost damage. When using methods such as wind machines and helicopters that produce an aggravating effect that can promote ice crystallization, it is critical to start the protection early before the plants are in a super cooled state.

The third category of frost protection focuses on *making the plants less susceptible to frost damage.* This is accomplished by making the plant more healthy and therefore more frost hardy, delaying bud break, and by retarding the formation of ice crystals. In this category there are copper sprays, nitrogen and chemical monitoring, and soil enhancements. Passive measures include late pruning, optimization of varieties and site selection.

Frost Protection Tools – Active Measures

Overhead Sprinklers - (In use since 1940's. Many systems installed in the early 60's.) Over tree irrigation is a common and effective method of frost control when used properly. The disadvantages of over tree irrigation are numerous and over tree irrigation can be a dangerous method that has the potential of causing serious damage when used incorrectly.

Prior to deciding on the use of over tree irrigation, the issues associated with proper use and causes for failure of the system should be considered and resolved.

Over tree irrigation can not be used for winter frost conditions, it is suitable only for spring and fall frosts as the temperatures during winter will cause the water lines and emitters to freeze.

Advantages
-Does not depend on inversion layer to be effective.
-Effective.
-Quiet.

Disadvantages
-High maintenance.
-Requires large amounts of water (50-80 gallons per minute/acre). The protection gained is directly proportionate to the amount of water applied to the canopy.
-Water should be applied uniformly per acre and not cycled.
-Can cause heavy ice load on plants.
-Must be started early due to a dip in the air temperature at start-up caused by evaporation.
-Can cause diseases from saturated soils, and the leaching of nutrients & other chemicals.
- Will cause more damage to the crop in the event of a system failure than if no frost protection was used.
-Overhead sprinklers for frost control can not be used with wind machines.
- Overhead sprinklers should not be used in any area that is wind prone, as wind and water will cause evaporation making the area colder than if no protection was used at all.

How They Work

Water emitted from the sprinkler heads encapsulates the crop by freezing around it. As more water falls on the encapsulating ice, the freezing of that water releases heat into the ice and keeps it just under 32 deg. F. This is cold enough to stay frozen and yet warm enough to coat the crop and act as a barrier from the surrounding colder air. This is what protects the crop from damage. Water that does not freeze on the crop because the spray pattern misses the target will either freeze in the air or fall to the ground where it will soak in or freeze on the ground surface. If the water freezes in the air or the ground surface, some heat will be released into the air but this is not a significant factor in the way that over tree irrigation protects crops. Since the protection is mainly derived from water freezing on the ice, the temperature of the water pumped through the system is not a factor unless it is too cold to travel to the sprinkler heads and freezes in the lines.

If constant application of sufficient water is not maintained, the temperature of the ice covering the plant tissues could fall to well below the surrounding air temperature. The ice that is covering the crops must remain wet on the outside with a steady drip of water off the ice. This will insure that there is a constant freezing of water going on and there is enough water freezing to release a sufficient amount of heat to keep the ice shield at a safe temperature. As more ice freezes on the crop and the ice shield grows larger, a larger volume of water is required to release enough heat to keep the larger mass of ice at a safe temperature. As long as water is dripping from the ice then there is enough water. If the mass of the ice builds up too large for the amount of water being applied, or the water supply is shut down, the ice will dry up and start to evaporate. This will cause the temperature of the ice to fall well below the temperature of the surrounding air and cause more severe damage than if there had been no frost protection at all.

When using micro sprinklers or pulsators, care must be given to the potential low temperatures in the protected area. A 15 gpm/acre micro sprinkler system operating at less than 27 deg. F is at risk of failure. The higher the volume of water in an over tree irrigation system, the lower the temperature that the system will protect against. The actual safe temperature is dependent not only on water volume but also the line sizes and types, emitter droplet size, and distance of water travel. Each system should be evaluated for the safe working temperature and compared with the actual known low temperatures in the area. For example, if there is a history of temperatures in the protected area going below 27 degrees F, a system with 15 gpm/acre or less is not suitable.

The less water volume in a system, the greater the risk of the water suddenly stopping due to its low temperature threshold. This will happen due to;

1. The lateral water lines freezing because there is not enough water mass and/or speed and turbulence through the lines to avoid freezing,
2. The emitter heads freeze at sufficient low temperature and clog the emitters shutting down water spray, and
3. Water that is emitted from the sprinkler heads freezes in the air before it reaches the ice on the crop because the mass of each droplet is too small. If the water freezes in the air, the heat is lost into the atmosphere and provides no benefit.

Over tree water must not be turned off until all the ice has melted and the sun has come up. If it is turned off earlier than this, there can be ice evaporation and damage.

The most critical point is the start up of an over tree irrigation system. During start up, the water spray will evaporate until the relative humidity of the surrounding air reaches 100%. In order for water to evaporate it must use heat from the air to vaporize. This causes the air temperature to drop substantially. This effect is known as 'evaporative dip', and every over tree irrigation system must contend with this. The procedure is to consult a chart that is supplied by the sprinkler manufacturer to determine the temperature dip at a specific relative humidity. The lower the relative humidity, the more evaporation will take place and the lower the temperature will drop. The sprinkler system must be turned on prior to the critical point where the current temperature minus the evaporative dip equals lethal temperature. If this point is missed, then the sprinkler system cannot be turned on. If it is, the crop may be damaged by the temperature dip. The crop will then be covered in ice as if everything was working fine, but in the morning after the ice melts, there are substantial losses due to cold damage that might not have occurred if the sprinkler system was left off.

Of course the standard problems of pump failure, or simply running out of water affects any type of irrigation system.

Most Effective Applications

Systems that use more than 50 GPM/Acre. Even micro sprinklers and pulsators should never be used with less than 30gpm/acre as the risk of failure is too high.
When used to overcome regional temperature deficits (advection freezes).
When used in conjunction with cold air drainage (Shur Farms Cold Air Drain®).

Least Effective Applications

Systems that are under 30gpm/acre for micros or under 50gpm/acre for full solid sets.
Wind prone hillsides and flat areas.
When used in conjunction with conventional wind machines.
Winter frost conditions

Under Tree Sprinklers and Flood Irrigation

Advantages

-Effective.
-Less risky than over tree sprinklers.
-Less water required than over tree sprinklers (25-45 gallons per minute per acre should be applied.)
-Lower application rates have the advantage of reducing water logging soils & leaching of nutrients & other chemicals.
-Good retention of energy.
-Quiet.

Disadvantages

-Requires inversion layer to be effective.
-Effectiveness depends on the amount of water converted to ice.
-Overall effect is small & confined to the area under the canopy.
-Misters and micro sprinklers work for under tree frost protection only if application rates and coverage are adequate. Small droplets and 'foggy' mist do not compensate for a low volume of water.

How They Work

Water emitted from the sprinkler heads or flooded along the ground freezes either in the air or on the ground. When the water changes state from a liquid to a solid (water to ice) 'latent heat of fusion' releases heat stored in the water. The surrounding air molecules are warmed by this release of heat and will rise due to thermal buoyancy to the point that they are the same temperature as the air around them. They will then 'stack' and begin to cool. Even after the warmed molecules are positioned at the point of stacking, they continue to cool and then descend back to the earth where they will be re-heated in the same manner and rise again. This process of continual up and down motion is convection and it is the same process that takes place with heaters. An artificial thermal ceiling is created and only the air molecules below that ceiling are heated and reheated to maintain a higher than natural temperature below the ceiling.

Unlike fuel heaters, the stacking point of the heated molecules when using under tree irrigation is never higher than the canopy of the crop. With fuel heaters the ceiling can be much higher. The temperature increase that can be attained is the temperature at the stacking height of the heated molecules, which is dependent on the volume of water applied and the conversion rate of the water to ice. Water that is emitted in too large of droplets may not freeze completely and soak into the ground. Water that soaks into the ground loses much of its frost protection ability, so when choosing emitters, droplet size and spray pattern consideration must be given to the expected low temperatures and soil conditions. When using flood irrigation, it is desirable to have the water slightly turbulent. Turbulence will increase the surface area that is exposed to freezing cold air and promote ice formation. It is ideal when there is a mixture of ice and water, or slush, on the ground with either flood irrigation or under tree sprinklers.

Even though under tree irrigation is associated with some evaporative cooling as with over tree irrigation, there is no danger to the crop because the colder air is already lower than the fruiting zone and cold air does not rise.

Failure of an under tree irrigation system due to loss of water will have no additional negative effects other than losing your frost protection. It is considered a 'safe' system.

Most Effective Applications

Under tree irrigation is useful in nearly all applications as long as the water is not applied to the fruiting surface.
Under tree irrigation should be applied as uniformly as possible throughout the protected area.
When used with cold air drainage (Shur Farms Cold air Drain)

Wind Machines

(First used in the 1920's in California, but not generally accepted until the 1940's-1950.)

Wind machines have gone through a long evolutionary process with a wide range in configuration and styles.

Fig 8 – Portable wind machine

Advantages

-Most beneficial in flat topography.
-Enhances the effects of heaters.
-Can work with under tree irrigation.
-Generally, the <u>maximum</u> gain in temperature is 45% of the difference between the temperature at the base of the wind machine and the center of the wind machine blade.
-Maximum benefit encompasses approx. 8% of the total coverage area.

Disadvantages

-Requires inversion layer to be effective.
-If the inversion layer is too high, stirring up the air with a rapidly rotating propeller is not going to bring warm air down to freezing fruit and can make the area colder.
-The effect varies throughout the protected area.
- Wind agitation can disrupt the plants ability to super cool.
-Noisy.
-Cannot be run in windy or advection conditions.

How They Work

Wind machines are wind mixers. Their main function is to push air from the warmer, higher stratus downward to mix with the colder air along the ground. A secondary effect is to push the cold air layer close to the ground out of the area or slightly upwards to mix. Most wind machines rotate 360deg. to cover a circular area.

Wind mixing does not assimilate cold air molecules into warm air molecules. There is little or no transfer of heat from one molecule to another. For a transfer of heat to take place there must be sufficient contact time between the warm molecules that are forced downward and the cold molecules along the ground. Due to the forces of thermal buoyancy, the warm molecules have a strong tendency to separate and rise back to their natural position above the cold molecules.

The continual rotation of a wind machine results in an end to the forced convection in a given spot and therefore, the maximum effect in any one area is temporary. When the warm air is blown down, there are warm molecules and cold molecules 'mixed' together. When the propeller moves and the wind stops blowing, then the warmer molecules will rise to their natural positions over the cold molecules until the propeller returns and re-mixes the air. This causes a continual rise and fall of temperatures throughout the protected area as the machine rotates.

The temperature increase due to wind mixing can be expressed as the 'sustained' temperature increase or the temperature 'flux'. Sustained increase is the average and flux is the temporary maximum rise in temperature. Because wind machines have a varying effect over the protected area, the sustained temperature increase should be used to determine the effectiveness of a system throughout the protected area and not the high temperature flux.

The maximum temperature increase that can be achieved with a wind machine is relative to the difference in temperature between the ground and the center of the wind machine blade. Depending on the distance from the wind machine, this difference is between 0% and 45%. It is therefore critical to locate wind machines in relation to the damage area with their effect, and also to avoid locating them in low pocket areas that accumulate cold air. Because wind machines are not designed to remove cold air accumulation, and cold air is continually added to accumulation areas throughout the night, cold air will build up around the wind machine and lessen the difference in temperature between the ground and the center of the wind machine blade rendering the wind machine less effective.

In fig 9, the distribution of warm air around the wind machine and the effects at given distances from the machine are shown. This chart assumes a flat area with no drift. In most instances the circumference is slightly more eccentric shaped. On slopes, the distance of the air throw is reduced on the uphill side and elongated on the downhill side. There is a circular band area that represents approximately 8% of the protected area that will have the greatest sustained and flux temperature increases.

It is also important that the wind machine blade be as high as possible to reach the warmest air available. Current technology limits this height to about 40 - 45 ft from the ground as the ability to propel warm air molecules to the ground from greater heights is not feasible. When locating a wind machine on hill, the distance to the protected area must be considered.

The greater the difference in temperature between the warm air molecules forced downward and the cold air molecules along the ground, the greater the force of thermal buoyancy. This results in more power being required to propel the warm air downward and less time warm and cold molecules will be mixed before separating and rising upward. This will result in higher temperature increase flux in relation to the sustained temperature increase.

In fig 10 the average temperature increase is charted. With a 10 deg. F temperature difference between the ground and the middle of the wind machine blade, the average temperature increase over 10 acres is 1 deg. F, but smaller areas have substantially higher average increases.

When placing a wind machine the best effect should be aligned with the damage area and care must taken to avoid a build up of cold air around the perimeter. Besides mixing warm air with cold air, a wind machine will push cold air around the perimeter of it area of influence. The cold air can accumulate there and submerge crops. The effect is a circle of damage around the perimeter of the protected area. Overlapping the affected circumference with other wind machines is one way to avoid the circle of damage that happens when a wind machine pushes cold air out of the area and it builds up around the edge of its area of influence.

Wind machines blowing uphill can obstruct cold air drainage resulting in accumulation and damage on the hill that would not have occurred otherwise. Another consideration is to avoid blowing cold air from one area to another or into a neighboring orchard. Position wind machines to blow cold air into a safe area.

Most Effective Applications

Flat areas and large open plain slopes.

Least Effective Applications

Frost pockets, gulleys and swales, cold air lakes.
Complex undulating topography.
Advection conditions and regional temperature deficits.
Crops less than 12" from the ground.
When used in conjunction with over tree irrigation.

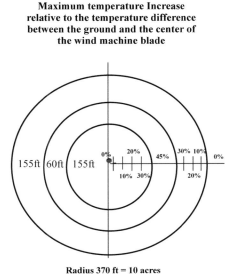

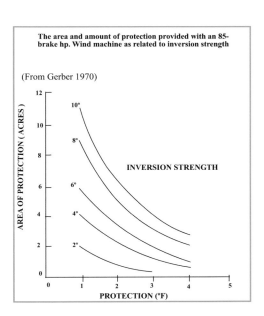

Fig 9 – Temperature flux chart   Fig. 10 - Average temperature increase chart

Shur Farms Cold Air Drain®

(Has been in use since 1990's)

Cold air removal technology is designed to eliminate micro climates by removing or preventing cold air accumulation.

Advantages

-Works in accumulation areas (severe & chronic frosty valleys, pockets, & swale areas).
- Can be used to control bud break
-Alters individual microclimates.
-Results are easily verifiable.
-More economical to run than wind machines or sprinklers.
-Environmentally friendly.
-Quiet.
-Removable at end of season.
-Low maintenance
-Uses passive measures (airflow dams & diversions) to maximize effectiveness
-Enhances the effects of over tree sprinklers, under tree irrigation, wind machines, and heaters.

Fig. 11 – Shur Farms Cold Air Drain®

Disadvantages

-Requires inversion layer to be effective.
-Cover crops must be maintained low to the ground to improve air flow.
-Obstructions to air flow must be removed

How They Work

The Shur Farms Cold Air Drain® lessens the temperature difference between the cold, frosty areas that accumulate cold air, and the areas that are naturally frost free. This is accomplished by removing the cold air that accumulates due to the lack of sufficient cold air drainage capacity. Under natural conditions, hillsides and other areas that are well drained do not experience frost damage because lethal cold air does not build up deep enough to submerge the crops. The coldest and heaviest air molecules along the ground are constantly drained off due to the gravitational pull causing katabatic air flow. As the coldest air is drained off, the warmer air layer from above becomes relatively the coldest and heaviest and drops down. As this layer is drained, then the next layer drops down and drains off. As long as the capacity of an area to drain off cold air exceeds the amount of cold air entering into the area, then there is sufficient drainage. An area that has sufficient drainage is termed naturally frost free, because of the inability of cold air to accumulate. With the exception of advection frost or a region wide temperature deficit where the entire region is submerged in lethal cold air, frost damage only occurs in areas where there is cold air accumulation. Even when there is a regional temperature deficit that is radiation in nature, the accumulation areas will suffer greater damage than the well drained areas.

The Shur Farms Cold Air Drain® removes accumulated cold air through a hydraulic process of selective extraction that mimics the natural process of cold air drainage.

The Shur Farms Cold Air Drain® extracts the coldest and densest molecules closest to the ground and forcibly drains them upwards out of the protected area. By removing the coldest air from along the ground, the warmer air layer above will move downward as it becomes relatively the coldest and heaviest layer. As this layer is drained, the next layer drops down.

The process used to move the cold air into the cold air drain is called 'Selective extraction of fluids of differing densities'. This is the same process that is used in removing water that accumulates in the fuel tank of an airplane and removing cream from milk.

In the airplane, fuel is lighter and less dense than water and so water in the fuel will separate and settle to the bottom of the tank. Opening the drain valve on the bottom of the tank will allow the water to flow down and out. Even though the fuel tank is the length of the wing, the water molecules along the bottom of the tank will flow horizontally for as long as necessary to reach the drain and flow out. After all of the water is drained out, the fuel becomes relatively the heaviest layer and moves downward. Since all of the water must be removed before any fuel can come out the drain, the pilot knows that when one drop of fuel comes out of the drain, all of the water is gone.

Instead of draining downward, the Shur Farms Cold Air Drain® forces the cold air upward through a specially designed wind tunnel. Depending on the application and whether the need is to capture a moving mass of cold air and direct it into the wind tunnel or to remove a static mass of cold air from a frost pocket, different designs of the Cold Air Drain® are used. All of the designs are used for the same basic purpose; the coldest air from along the ground is pulled into the bottom of the machine and ejected upwards. As the cold air is ejected upward it crosses through the warmer air layers. As the cold air molecules pass through warmer air, mixing of cold and warm molecules begins. This is termed 'ascendant mixing' because the mixing is caused when the ascending jet stream of colder air crosses the warmer air layers. Warm air is drawn into the jet stream and mixed with the cold air. Transfer of heat from the warmer air molecules into the colder air molecules warms the jet stream, lessening its density and making the molecules lighter.

When the jet stream reaches its dissipation height from the ground, approximately 300ft, the temperature and density of the air molecules in the jet stream is approximately the same as the surrounding temperature. At this point the air disperses horizontally into the upper inversion. The cold air that is removed from the ground has changed density and temperature through the assimilation of warm air into the jet stream. It is permanently removed and cannot fall back down to earth.

Cold air removal can be used to drain accumulation areas or to prevent cold air from entering into an area and thus avoid accumulation altogether. There are seven major ways that cold air can accumulate which will be discussed in a later chapter. Draining cold air cannot cause additional damage under any circumstances, although it is effective in preventing frost only under radiation conditions. Only under radiation conditions does atmospheric stratification exist. Under advection conditions, the machine would revert to non-selective extraction and remove air from the area immediately surrounding the machine. Since under advection conditions the air is the same temperature throughout the zone, there is no detriment to this effect, and since advection conditions can quickly turn into radiation frost with the clearing of the skies (and back to advection), it is beneficial to leave the cold air drain running. It is considered a safe method.

Cold Air Drain® in Conjunction with Other Active Frost Protection

Cold air drainage is compatible with all other forms of frost protection. The effect is to lower the ceiling thereby enhancing their effects by simulating a stronger inversion. Most methods of frost protection rely on an inversion to be effective and the stronger the inversion, the greater the effect. Care should be taken to create a 'sink' area around the Cold Air Drain® machine of 20ft around the perimeter where the air is undisturbed. To create a sink area, the Cold Air Drain® should be located at least 325 ft from a wind machine and 20ft from any sprinklers or heaters. This allows the cold air to settle and reach maximum stratification prior to entering the machine.

Most Effective Applications

Cold air pockets, swales and gulleys, lessening slopes, converging air currents, slopes and other areas of cold air accumulation.
Perimeter barrier to avoid cold air entrance along an uphill side.

Least Effective Applications

Regional climate deficits and advection frost conditions.
Crops less than 12" from the ground.
Row crops where the rows are perpendicular to the slope.

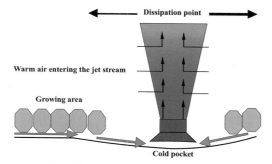

Fig. 12- Cold Air Drain® in a static air mass

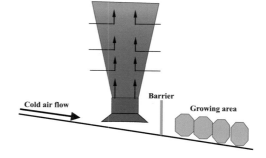

Fig 13 – Cold Air Drain® in a dynamic flow

Heaters

(Fossil fueled heater technology was developed in the early 1900's-1920's.)

Fig 14 – Return stack heater

Advantages

-Heaters throw off some heat.
-Work best when used with wind machines and in accumulation areas or static air mass.

Disadvantages

-Expensive to operate. (Oil requires 30-40 heaters/acre. Propane requires approx.70 heaters/acre. A heater usually burns between .5 and 1.2 gallons of fuel per hour.)
-Labor intensive
-Air pollution and environmental issues may be problematic.
-Between 75-85% of heat from conventional oil or propane heaters can be lost through radiation into the sky.

How They Work

Heaters warm the surrounding air. The heated air becomes lighter and less dense than the surrounding colder air causing it to rise. The heated air begins to cool as it ascends and at some point above the ground, it is the same temperature and density as the surrounding air. The greater the difference in temperature between the heated air and the surrounding cold air, the faster and higher the heated air will rise. 'Thermal buoyancy' is the term used to describe the phenomenon of warm air rising, and the greater the difference in temperature between the cold air and warm air, the more 'buoyant' the warm air becomes.

When the heated air rises to the point where it is the same temperature as the surrounding air, it becomes 'neutrally buoyant' and will stop rising. At this point the thermal boundary line that is created is known as the 'ceiling'. Since a molecule of a specific temperature, for example 30deg. F will not push away another molecule of the same temperature and density, heated molecules of the same temperature will begin to 'stack' under each other. As more air is heated, it will eventually stack under the ceiling until the temperature in the entire area is equalized.

As the heated air molecules begin to cool, they become less buoyant and start to descend back towards the ground. Other heated molecules take their place. The cooling molecules will fall until they are close enough to the heat source to be re-heated and rise again. This process is called 'convection'. Only the molecules that are stacked below the ceiling are heated thus avoiding the total loss of heat into the atmosphere.

The best heaters operated correctly are 75% -85% inefficient. This means that 75-85% of the fuel burned to create heat is lost to the atmosphere, doing no work and providing no benefit. In reality no other process of any kind in any industry would tolerate such inefficiency.

The ceiling height can be controlled by modifying the amount of heat or heat concentration coming from the heat source. A more indirect heat – or radiant heat – will impart the same amount of heat energy (BTU's) over a larger area, while turning the heaters down and burning less fuel will provide less BTU's. Either approach will provide less temperature difference between the heated molecules and the natural molecules resulting in less thermal buoyancy. This will result in a lower ceiling or 'stacking' height. In most cases the ideal ceiling height is just slightly above the canopy. Using more heaters, each with less fuel consumption will have the same effect as fewer heaters that are more radiant.

When a molecule is super heated with a direct non-radiant heat source such as a bonfire, then the heated molecule will rise to much higher level before it stacks and may never have the chance to convection back down into the protected area before the damage is done or the sun comes up. In these cases, the grower will only have succeeded in heating the atmosphere above the orchard, making no impact on the temperature inside the orchard.

The proper method for using heaters is to start with a low heat volume in order to build a 'ceiling' and to quickly complete the convection process. No real protection takes place until this convection process is complete. When the temperature below the ceiling is equalized, the heat can be slowly turned up in order to raise the ceiling height.

The limit of the temperature rise inside an orchard is the temperature at the ceiling. For instance, if the temperature at 15ft above the ground is 3 deg. F higher than at the ground, and this is where the ceiling is set, then the maximum rise in temperature inside the orchard will be 3 deg. F. In order to achieve a higher temperature rise, then the ceiling must be raised which would require significantly more fuel in relation to the temperature increase.

Since the thermal gradient (the temperature rise at a given height increment) spreads out as the height from the ground increases (for instance there may be a 3 deg F. rise from the ground to 15ft, but the next 15 ft may only rise 1.5 deg. F, and the next 15ft after that may rise only 1/2 deg F), fuel consumption rises exponentially to the rise in temperature in the protected area. Due to time constraints, there is a maximum height that the ceiling can be established at in order to complete the convection process before damage occurs to the crop. The earlier the heaters are started, the higher the ceiling can be. The limit to the temperature increase with heaters in a protected area is determined by the strength of the inversion and the cost of fuel that the grower is willing to endure.

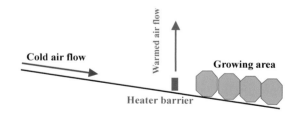

Fig. 14 – Perimeter heater barrier

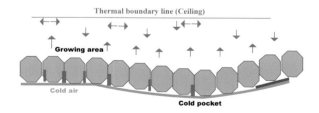

Fig 15- Heater convection currents

Most Effective Applications

Heaters work best in accumulation areas and areas with a non-moving (static) mass of air such as flat ground, cold pockets, and cold air lakes.

Least Effective Applications

Areas with dynamic air flow such as hillsides and swales. In these areas, the heat released into the atmosphere is quickly replaced be fresh cold air entering the area. This is the same as trying to heat a river; all of the water that is heated is quickly replaced by fresh cold water.

Uphill Perimeter boundaries - Heaters that are positioned at a close distance to each other along an uphill perimeter boundary to create a thermal barrier will not provide heated air into the protected area down hill. The heat from the heaters will not flow downward into the orchard – heat only goes up, never down. The effect is to heat the cold air moving downhill from outside the orchard (due to katabatic flow). When the cold air molecules come in contact with the heater barrier, the cold air that would normally flow into the orchard will be heated and rise thus blocking its entrance into the orchard. Since it is the cold air generated outside and upstream of the orchard that causes most or all of the damage inside the orchard, the avoidance of cold air entrance does offer substantial protection against frost damage, however creating a barrier with heaters is inefficient and only marginally effective. Due to large gaps of heat close to the ground between the heaters cold air can 'leak' through the barrier and the high cost of this method makes it by far the least efficient way to block cold air from entering an orchard.

Heaters in Conjunction with Wind Machines

When wind machines and heaters are used together they work in a completely different way than when either wind machines or heaters are used separately. Wind machines and heaters together is essentially forced air heating. It is possible to achieve a high temperature increase because the wind machines will capture and recirculate more of the heat produced by the heaters through forced convection than is achieved through natural convection. Wind from the wind machines act as the ceiling and forced convection is faster acting than natural convection.

When using wind machines and heaters during an advection freeze, the wind machines will blow colder air down to mix with the heat produced by the heaters. Under advection conditions, it is essential that the heaters be positioned to provide complete coverage of the area and that there is sufficient heat applied to overcome the temperature loss due to blowing colder air downward.

Fog Systems

Advantages

-Effective
-Low water usage
-Low danger of soil saturation
-Quiet

Disadvantages

-Drift of fog is difficult to contain.
-Fog systems in the US are rare due to the liability issue of fog drifting onto highways and causing accidents.

How They Work

There are two major ways that fog works to prevent frost. First, the ground loses heat through long wave radiation. These radiation waves are approximately 10-15 microns in diameter. Natural fog particles are the same size as these radiation waves, and so the fog will absorb and reflect the heat lost from the ground preventing it from escaping into the atmosphere. This will keep the ground warmer, slowing the cooling of the lower layer of the atmosphere.

The second effect is that fog will eliminate atmospheric stratification and the associated temperature differences within the fog zone, equalizing the temperature throughout the fog zone eliminating any micro climates.

Since fog is affected by gravity, it will flow in the same manner as cold air along the ground. It will find the coldest spots and get into hard to reach places. Because fog is affected by gravity, it is difficult to maintain a deep layer. Fog tends to compress and become denser when more is added instead of maintaining its density and building up a deeper layer. For this reason, the best use for fog is to retard heat loss from the ground.

Artificial fog that is indistinguishable from natural fog can be made by forcing water through an orifice at extremely high pressures to achieve the correct particle size. A well distributed fog bank will have about the same effect as an under tree irrigation system with less than 10% of the water usage.

Most Effective Applications

Frost pockets, static accumulation areas and flat areas.

Least Effective Applications

Windy areas.

Helicopters

Advantages

- Helicopters are a form of wind machine
- No water usage

Disadvantages

- Expensive
- Dangerous
- Can cause severe damage if called too late
- Unreliable due to too many variables

How They Work

Helicopters are a form of wind machine. They push warm air from above to mix with the colder air along the ground, same as a conventional wind machine. Sometimes the grower will install temperature sensitive lights throughout the protected area that will go on when the temperature falls to a critical level. The helicopter pilot will fly from light to light to 'put them out' by raising the temperature.

One helicopter is normally used to protect between 25-50 acres. The capacity of a single helicopter is dependent on the weight and size. The heavier they are, the more power can be applied and the more air that is pushed down.

Most Effective Applications

Flat areas and large open plain slopes.

Least Effective Applications

Overall, helicopters are not reliable and way too expensive to be considered a primary source of frost protection. Delays in arrival, pilot skill, helicopter types, weights and styles, and other variables prevent helicopters from being an effective method of frost protection.

Chemical Sprays

Advantages

- Inexpensive
- Easily applied

Disadvantages

- Most studies show that chemical sprays of zinc & copper show no measurable benefits.
- Sprays to eliminate ice nucleating bacteria are ineffective because of the natural abundance of ice nucleators already in the bark & stems, and the apparent ability of ice to nucleate around dead bacteria.

How They Work

Copper based sprays are designed to reduce the numbers of ice nucleating bacteria. Copper is a bacteria stat, which means that it will inhibit the introduction of new bacteria and as the old ones die off, the population decreases. The theory is that after several days from application, there will be lower numbers of ice nucleating bacteria and less risk of ice crystallization in the plant cells. Because of the lower risk of ice crystallization, the plants should be able to super cool to lower levels.

New technology of introducing competing bacteria to lessen the presence of ice nucleating bacteria has shown some promise, but is not currently available due to issues related to genetic engineering.[13]

Other types of sprays such as polymers and elastics are designed to seal the plant tissues to avoid transpiration of water, keeping the plant from 'freeze drying'. By keeping the plant hydrated, it is assumed that the plant will be able to super cool to lower temperatures before ice crystallization begins.

Most Effective Applications

Neither copper based nor polymer sprays have shown a consistent ability to enhance the natural super cooling of the plants.[14] Copper sprays are effective in controlling some fungus and bacteria that can cause damage to the plant. This makes for a healthier plant and a healthier plant is more frost resistant than a sicker one.

---

[13] **The use of genetically engineered bacteria to control frost on strawberries and potatoes. Whatever happened to all of that research?** R.M. Skirvin, E. Kohler, H. Steiner, D. Ayers, A Laughnan, M.A Norton, M. Warmund Univ. of Illinois, Dept of Natural Resources and Enviornmental sciences, Univ. of Missouri, Dept of Horticulture

[14]**Spring Frost Control**   North Carolina Winegrape Growers Guide, NC State Univ. cited Sugar et al, 2003

Burning Debris

Advantages

- Releases some heat

Disadvantages

- Environmentally disastrous.
- These fires provide little radiant heat, and smoke does not provide any frost protection.

How It Works

Burning debris is a form of heater.

Most Effective Applications

Setting up and burning many small fires.

Least Effective Applications

Burning large, hot fires with lots of smoke.

Frost Protection Tools – Passive Measures

Passive frost protection methods are those that require no outside energy such as burning fuel or electricity. Passive measures work by helping to create a natural environment that promotes health and optimum growing conditions for the crop.

Site and Varieties Selection

Advantages

- Can be integrated into existing management practices at little or no cost.
- Environmentally sustainable practice.
- Optimizes variety suitability within a growing area.

Disadvantages

- Data logger study may be required to determine the temperature correlation between microclimates in a growing area.
- A study may have high costs.
- Should be done prior to planting.

How It Works

Site selection is the most effective and efficient method of any frost protection. Choosing a site that avoids frost conditions by being naturally well drained, and matching the crop to the regional climate, soil and other conditions will ultimately have the best return on investment of any frost protection method.

The larger the growing area, the more likely it is that there will be areas that accumulate cold air. In rolling hills and undulated topography there will always be areas that experience cold air accumulation as the higher spots drain themselves. The drained cold air will eventually either be obstructed or enter into an area faster than it can drain out. Areas that are at risk for cold air accumulation are sometimes not planted wasting valuable growing area.

Barriers and Diversions

Advantages

- No fuel required and does not need to be turned on or off.
- Effective when used properly.
- Environmentally sustainable practice.

Disadvantages

-The proper design and placement of barriers and diversions is essential.
- Barriers and diversions have the potential to cause serious damage if not used properly.
- May have high installation and acquisition costs.

How They Work

Barriers block or divert cold air that moves along the ground.

The use of barriers to accomplish manipulation (either as a barrier or diversion) of cold air currents is the most effective and powerful passive frost protection method after site selection. The existence of heavy vegetation at the low end of a vineyard or orchard can result in the obstruction of cold air drainage and may be the cause of severe frost damage. A similar obstruction along the high side perimeter can block cold air entrance into the vineyard and result in an improved situation for the downstream vineyard, but have an equal negative effect on another area where the cold air backs up or is diverted to. Barriers can be used to create dams and obstructions to air flow as well as diversions that will deflect the cold air flow out of the protected area and into a safer location. Barriers have the potential to create a substantial impact, but barriers alone do not remove the cold air so that leaves the question of where is the cold air going? Barriers are most effective when used in conjunction with cold air drainage.

Air flow obstruction or diversion can be accomplished by natural or artificial methods, such as heavy vegetation, soil berms or non permeable artificial barriers. Buildings and road embankments also serve as barriers, and heaters can create a thermal barrier blocking the entrance of cold air flows. Wind machines blowing uphill can create a barrier to cold air flowing off the hill.

The existence of a barrier at the low end of an orchard can result in the obstruction of cold air drainage, causing cold air accumulation resulting in severe frost damage, but in another similar circumstance where there is cold air buildup from below, the same barrier might act to block the entrance of cold air into the orchard.

The existence of a barrier at the high side of an orchard can block cold air entrance into the orchard. By creating an obstruction to the air flow and slowing the incursion of cold air into the orchard there will be an improved situation for the downstream orchard. If there is an orchard upstream of the barrier, and since a barrier does not remove cold air, there will be an equal negative effect of cold air buildup to crops being grown above the barrier as a result of cold air accumulation. Eventually, cold air will breach the barrier, flowing over the top, and finding gaps and other openings. Cold air will build to a height above the barrier that is determined by the height of the barrier, angle of approach and the viscosity of the air. When the buildup reaches its maximum containment height, air that flows over the top will mix with warmer air aloft and continue to move downhill.

An obstruction built to intercept cold air flows at an angle in an effort to deflect and divert the cold air flows to a safe area must consider the viscosity and low momentum of katabatic air flows. Mostly, cold air flows see these types of diversions as barriers and will tend to build up behind and around them. Diversion barriers must be designed with the proper height and angle of interception to keep air flows moving.

Barriers of any sort should never be used or modified without the benefit of an air flow model and analysis of the effects of such a barrier or diversion. Barriers have the potential to create a substantial impact both positive and negative.

Barriers alone do not remove cold air so that leaves the question: Where is the cold air going to go?

Most Effective Applications

Upstream the orchard to block out cold air flows from above.
When used in conjunction with cold air drainage.
Diverting lateral cold air incursion

Least Effective Applications

When used without cold air drainage.
Improper placement

Soil Management

Advantages

- Can be integrated into existing management practices at little or no cost.
- Environmentally sustainable practice.
- Can raise night time temperatures 1 - 2 deg. F

Disadvantages
- None

How It Works

Cultivation- Loose soil will lose heat during a radiation frost event faster than dense, packed soil. Disking loosens the soil and provides more surface area to transfer heat out of the soil. While disked soil also absorbs more heat during the day for the same reason, the additional heat absorbed is quickly lost after the sun goes down.

Water application- Darker soils will absorb more heat than light colored soils and soil that is moist along the top 12" of the surface will retain more heat than dry soils during a radiation frost event. Applying water to the soil surface during the day will darken the soil and allow for more heat to be absorbed. Keeping the top 10"-12" of the soil surface moist at night will retard heat loss and keep the soil warmer.

Soils should be mowed and left hard packed instead of disked or tilled during the frost season and kept moist to a depth of 10" -12" to help retard heat loss from the ground during the night. Since it is the ground that gets cold and cools the air that causes frost damage, slowing the cooling of the soil surface will slow the cooling of the air.

Most Effective Applications
All areas.

Least Effective Applications
None.

Cover Crop Management

Advantages

- Can be integrated into existing management practices at little or no cost.
- Environmentally sustainable practice.
- Can raise night time temperatures in some cases up to 4 deg. F

<u>Disadvantages</u>

- May impact soil nutrients and natural pest and weed control.

<u>How It Works</u>

During the frost season cover crops should be kept to a minimum in areas that are prone to frost. The recommended maximum height is less than 3" from the ground. During the day, a thick cover crop will prevent sunlight from hitting the ground. This will retard heat absorption into the ground making the ground colder at sundown. An area with thick cover crop can be 1-2 deg. F colder during the night than the same area with the cover crop removed.

On a slope, cover crop will obstruct cold air drainage, creating an accumulation of cold air on the slope and causing frost damage. A slope with 12" high cover crop can be as much as 4 deg. F colder at 48" from the ground than the same slope with cover crop kept to less than 2 ½" high. On slopes, cover crop is never an option during frost season and must be kept to a minimum.

There are two ways that cover crop can be beneficial for frost protection. The first is with some under tree irrigation systems. Water from the sprinklers will be trapped in the cover crop, preventing it from soaking into the ground and allowing it to freeze releasing heat into the protected area. If the cover crop is on flat ground in an area that is shielded from the sun by a thick canopy or other obstruction then there is no detriment, otherwise the loss of heat into the ground from the sun far outweighs the benefit of additional water that will freeze.

The second is that under rare circumstances, usually after several days of cloud cover, the soil will retain more heat with a thick cover crop during the day. This happens when the daytime conditions, cold and cloudy, are such that the soil does not absorb heat but instead gets colder. The cover crop acts as a blanket. In all other instances, cover crop is not beneficial for frost protection. Water is best managed by optimizing the spray pattern and droplet size, and the advantages under the second circumstance are minimal.

<u>Most Effective Applications (for removing the cover crop)</u>

Remove cover crop in sun bathed areas and on any slope.

<u>Least Effective Applications</u>

Cover crop management will have the least importance in flat areas under a heavy canopy and other sun obstructed areas.

Late Pruning, Chemical and Other Dormancy Extending Methods

Advantages

- Can be integrated into existing management practices at little or no cost.
- Environmentally sustainable practice.
- Slows bud break and extends dormancy

Disadvantages

- May impact harvest dates

How It Works

Late pruning, chemicals and other methods that can delay bud break will keep the plant dormant until later in the season when the risk of frost decreases. Even a few days of extended dormancy can significantly decrease the risk of frost. Check with your orchardist or viticulturist to see which methods may be suitable to your specific varieties.

Methods of Frost Protection That Should Never Be Used Together

Often, it is possible and even desirable to combine frost protection methods to increase the protection factor. In some instances, more than one method will work together synergistically and produce a greater result than the sum of each one working alone.

In some cases, methods are not compatible and if used together will cause damage where no damage would have otherwise occurred.

1. Wind machines and over tree/ over vine water irrigation- These methods should never be used together because the wind will cause evaporation. This will produce an evaporative cooling effect, driving the air temperature down. Even if the natural air temperature would not have fallen below the crop's critical level, damage may occur that would not have occurred if only wind machines or only water were used alone. Over tree water is also not suitable for any area that is not wind protected such as hillsides and flat plains with a history of wind.

2. Helicopters and over tree/ over vine water – These methods should never be used together for the same reasons as wind machines and over tree water. Helicopters are a version of conventional wind machines.

The use of helicopters is not recommended under any conditions due to the high cost of operation in relation to the protection afforded and the inherent danger associated with flying helicopters at low altitude during the night. In many cases, the high cost will delay the implementation of the helicopters until well after the crop is in a super cooled state. When this happens, the air movement can stop super cooling and actually promote ice crystallization causing frost damage when none may have occurred naturally. It is not uncommon for helicopters to cause damage in this way.

3.   Heated water for under tree irrigation - Water used for under tree or under vine frost protection should never be heated. The largest benefit of under tree water is realized when the water freezes. The conversion from water to ice is the main way that this method releases heat into the air and protects the crop. Any water that is not converted to ice and subsequently soaks into the ground is lost as far as any protection value is concerned.

Water that is heated may not cool sufficiently to freeze and then soak into the ground, losing the bulk of its potential to heat the air.

Methods of Frost Protection That Should Not be Used in Certain Situations

Different frost protection methods protect in different ways. While some are well suited to solving a particular problem, they may not be suited at all to another problem. Active and passive measures of frost protection create an action that modifies either the causes of cold air accumulation or the effects. Where there is an action, there is a re-action which can cause an unwanted effect.

There are three frost protection methods that can be termed risk free, meaning that these methods, no matter how or where they are used, will not cause frost damage. These methods are 1. Heaters, 2. Under tree irrigation and 3. Shur Farms Cold Air Drain®. If these methods are not used properly, then the effects will be diminished, but there is never a negative effect even if the system fails completely.

The following methods should not be used in these situations as damage from the frost protection system itself may occur;

1.   Over tree irrigation on hillsides- Hillsides are not usually wind protected. Over tree irrigation should only be used in wind protected areas because wind blowing on water causes the water to evaporate. The evaporation of the water on the hillsides will cause the air temperature and the temperature of the ice shield around the plant to drop. Frost damage can occur as a result of the water application in an area that may not have otherwise had damage.

2. Wind machines placed at the bottom of a hillside- A wind machine blowing uphill on a hillside can create an obstruction to the natural drainage of cold air off a slope. This will cause cold air to back up on the hillside.

3. Wind machines or helicopters during cloudy or windy conditions- Under cloudy or windy conditions there is no inversion. When there is no inversion, the temperature drops as the distance from the ground increases. The wind machine or helicopter will blow colder air down into the orchard causing damage. Wind machines also cannot be used in windy conditions because the propeller will become unbalanced causing severe damage to the propeller and to the machine.

4. Wind machines or helicopters after the crops are in a super cooled state- When the crops are already below their critical temperature, blowing wind over them can cause ice crystallization to begin. Wind machines and helicopters must be started before the crop goes into a super cooled state.

5. Wind machines or helicopters in conjunction with over tree irrigation- Wind and water will cause evaporative cooling. The effect of evaporative cooling will be to further cool the air temperature.

6. Over tree irrigation when the temperature and relative humidity falls below the safe level in order to recover from the evaporative dip- When over tree irrigation is turned on, the water will evaporate until the relative humidity in the surrounding air reaches 100% or fully saturated. This evaporation will cause a drop in temperature. An irrigation frost protection chart should be used to determine a safe temperature to turn on, usually 3-7 deg. F above the critical temperature for the crop. The safe temperature varies depending on relative humidity and volume of the system. If the temperature is too low to recover from the evaporative dip, over tree irrigation must not be turned on.

7. Over tree micro sprinklers in areas where the temperature is known to fall below 27 deg. F. - Over tree micro sprinklers at 15 GPM / Acre have an increased risk of failure at temperatures below 27 deg. F. There are three ways for micro sprinklers to fail, 1. Water in the supply lines will freeze, 2. The emitters will freeze and clog, and 3. The water droplets from the emitters will freeze in the air before they land on the ice surrounding the buds. All of these failures are caused by the low volume of water in the system. For increased protection, more water per acre must be introduced.

8. Heaters in dynamic flows. Any heat that is applied in a swale, cold air river or other dynamic air flow will be quickly replaced with fresh cold air from upstream.

Chapter 6: Cold Air Accumulation

Patterns and Analysis

The underlying cause of frost damage is cold air accumulation.[15] Where cold air accumulates, a micro climate is created that may be several degrees colder than the surrounding area. Hillsides, slopes and areas that are well drained are naturally frost free because these areas do not accumulate cold air.

Cold air accumulation can occur in a static or a dynamic mass of cold air. A static mass of cold air is where there is restricted cold air movement caused by obstructions or a low slope angle causing cold air to be trapped and pool up. When the depth of the lethal cold air is high enough to submerge the crops, frost damage can occur. A dynamic mass of cold air is where there is a down slope or katabic flow. A flowing mass of cold air can build up because cold air is continually added to the flow causing the depth of the thermal boundary line to rise. When the depth of the cold air mass gets deep enough to submerge the crop in lethal cold air, frost damage can occur.

When cold air builds up behind an obstruction, the obstruction is said to be a 'cold air dam'. Behind the dam or in a low 'hole', a cold air lake will form. Damage patterns in cold air lakes resemble the water line in a lake and follow a topographic gradient line. Damage starts in the area where the accumulation of cold air starts, which is the coldest spot in the basin and gets less severe as the elevation increases. The spot where the accumulation of cold air starts is not necessarily the absolute lowest spot in the basin. In some cases, the absolute lowest spot in the basin will also have some drainage capacity. This can lower the thermal boundary line resulting in higher temperatures and less frost damage than in a higher spot where there is a convergence of cold air streamlines even when that is not the absolute lowest spot.

Cold air needs at least 1/2 % slope grade in order for katabic air movement to occur. Where little or no movement of cold air along the ground exists, there can be a cold air buildup from ground cooling directly below the area. In these flat areas, the damage pattern will be even throughout the entire area, with the most serious damage occurring in the lower parts of the trees and damage getting less severe higher up.

---

[15] **Frost Risk Mapping for Landscape Planning: A Methodology**
G. P. Laughlin ~ and J. D. Kalma 2
With 8 Figures
Received March 25, 1988
Revised September 21, 1989

Cold air building up below the protected area can flood into the area. The cause of the cold air buildup below can be either a static or dynamic air mass. When the build up is caused by a static air mass building below the protected area, the damage pattern will resemble a cold air lake where the damage will be most severe in the lowest area and less severe as the elevation increases and along a topographic gradient line. When the buildup is caused by a dynamic flow, most commonly along a creek or drainage, the damage pattern will maintain a relatively constant distance into the orchard from the drainage, crossing gradient lines.

In distinguishing between a flooded area and a cold air lake, in flooded areas the temperature of the basin or drainage below the orchard will be colder than in the orchard and it will get cold first. In a cold air lake, the area below the obstruction or dam will remain warmer than in the orchard. Along a drainage, if the damage is caused by the drainage filling with cold air and flooding over its banks, then the rows of trees nearest the drainage will experience heavier damage than the inside rows because the cold air enters here and submerges the crops along the drainage first.

Cold air accumulation caused by less than 100% obstruction of the cold air that would normally go into a drainage will result in the first couple of rows of trees closest to the drainage having less damage than the inside rows. This is because the drainage is warmer than the crop area and the first row or two is closer to the warm air. There tends to be some drainage from the crop area into the drainage from barrier leakage which causes the thermal boundary layer to thin out along the outside rows. This same phenomenon occurs to the first few rows of crop planted along roads and open areas where drainage is enhanced away from the crop.

Accumulation in a dynamic flow is caused by cold air streamlines building up due to more cold air flowing into a sloped area than can be naturally drained off. Cold air will flow along a sloped ground and into gullys and drainages creating a cold air river. Where two or more of these cold air rivers merge, there is a convergence of air currents that can cause a deepening of the cold air mass if the downstream drainage capacity is not increased to handle the additional mass.

Accumulation of a moving air mass occurring on a plain slope where the slope angle remains constant while the cold air mass increases, eventually raises the thermal boundary layer high enough to submerge plant tissue. If the slope angle gets less while the cold air mass remains the same, a slowdown in the speed of the air flow will occur at the point of the lessening slope even if the overall slope angle is sufficient to drain the entering cold air and keep the thermal boundary below crop level. This slowdown will cause the cold air to basically pile on top of itself creating accumulation of cold air and raising the thermal boundary layer.

Converging streamlines of cold air at opposing angles or near opposing angles, such as when two or more cold air rivers come together from opposite sides of a basin, can cause a build up of cold air due to the air flow currents acting on each other creating eddy currents. Eddy currents act chaotically and produce their own obstruction. The damage pattern can resemble a cold air lake or pool of cold air without the structure of a constraining obstruction or of being inside a gradient line.

These are the seven major ways that cold air will accumulate in the two major categories of Static air mass and Dynamic flow. There may be multiple accumulation processes going on in any given area at the same time.

Static Air Mass	**Dynamic Flow**
Cold air lake • Flooding	Cold air river • Plain slope
Flat area accumulation	Converging air currents • Lessening slope

Accumulation Process #1 - Cold Air Lake

Cold air lakes exist where there is a static mass of air that builds up or pools in a specific area.

One way a cold air lake is created is when there is insufficient drainage of cold air caused by an obstruction such as a canal bank or raised road embankment, heavy vegetation or buildings obstruction. This will cause cold air to fill behind the obstruction. Cold air lakes will also form in low pockets and holes in flat ground.

Because of the high viscosity of cold air, under certain conditions such as flat area accumulation and converging air currents, a cold air mass of sufficient depth to cause frost damage can form where there are no physical barriers of containment. In effect, this is a simulated cold air lake.

The coldest spot in a cold air lake is the point where the cold air starts to accumulate and is commonly, but not always, the very lowest spot.

Fig 15 – the road embankment is a barrier to cold air drainage.

In the above photo, the road embankment is a barrier to cold air flow. As cold air flows from the higher areas and is stopped by the raised road embankment, it will back up and create a cold air lake. The height of the embankment is not the limiting factor for the height of the cold air mass. Depending on the angle of approach of the slope to the obstruction and the temperature and viscosity of the air, the cold air mass can build to 2 to 3 times the height of the obstruction and then back into the orchard.

The height of the thermal boundary line can be considered in relative terms as the separation point of the lethal temperature cold air mass and non lethal temperature for the protected crop, or in absolute terms as the height that cold air molecules entering into the basin displace the existing air mass upwardly.

As cold air enters the basin and builds up, the warmer air in the basin is displaced upwards. The lower the angle of approach to the obstruction, the deeper the cold air will buildup in relation to the height of the barrier before finally flowing over the barrier.

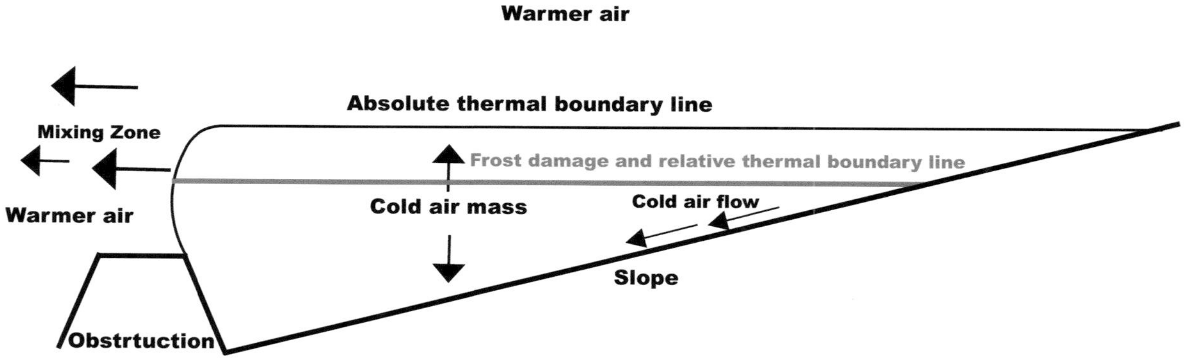

Fig 16– A low slope angle to the obstruction will allow a buildup of cold air that is higher than the height of the obstruction, not all of which may be lethal to the crop.

Cold air that flows over an obstruction will mix with the warmer air on the other side. This mixing process will raise the temperature and result in higher temperatures on the downside of the obstruction.

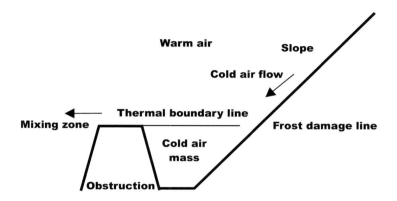

Fig 17- A steep slope angle to the obstruction will build up a cold air mass that is at least the height of the obstruction itself before flowing over.

Fig 11 – A vegetation obstruction to cold air flow.

Fig 18– A low spot that is creating a cold air lake.

A low area or frost pocket as shown in fig. 18 is created when higher areas surround a low spot. Cold air flows in but cannot flow out due to the lack of drainage. In some instances, there may be some natural drainage but if the drainage is not sufficient then cold air will accumulate and build up in the pocket.

Accumulation Process #2 – Cold Air Rivers

A cold air river is a concentrated streamline of cold air that is created when cold air flows into a gulley or swale and continues downhill. This is a dynamic (moving) mass of cold air that is contained within swale boundaries and the mass of cold air can get wider and deeper as it flows downhill. The increase in the mass of cold air downstream is due to the addition of cold air that enters the gully from the surrounding area. There will be a difference in temperature between the air inside of the swale and the air outside. The air inside the swale can be several degrees colder because the coldest and heaviest molecules will flow into the gully, displacing the warmer air layers. Frost damage that 'snakes' its way through an orchard is an indication of a concentrated streamline of cold air.

There are similarities in the development of a cold air river to the development of a water river. The moving cold air is referred to as a streamline. Like any water shed area there may be several small stream lines that converge into each other and then into a main drainage or main stream. As the smaller streamlines merge with each other, and eventually empty in to the main drainage channel, the depth of the cold air mass can rise and the width can widen due to the increasing volume of cold air.

Under dynamic flow conditions, the absolute thermal boundary layer is defined as the separation point of the dynamic (moving) and static (non-moving) air molecules. Above the absolute thermal boundary line the molecules are too light and too far from the ground to be affected by gravity and will not move downhill. The relative thermal boundary line is the point at which the lethal cold air is separated from the non-lethal temperature cold air. Below the thermal boundary line, the molecules are affected by gravity and will flow downhill. Typical damage patterns in a cold air river start at the point where the relative thermal boundary line submerges the plant tissue.

If the relative thermal boundary line is higher than the absolute thermal boundary, then this indicates a regional temperature deficit or an advection frost.

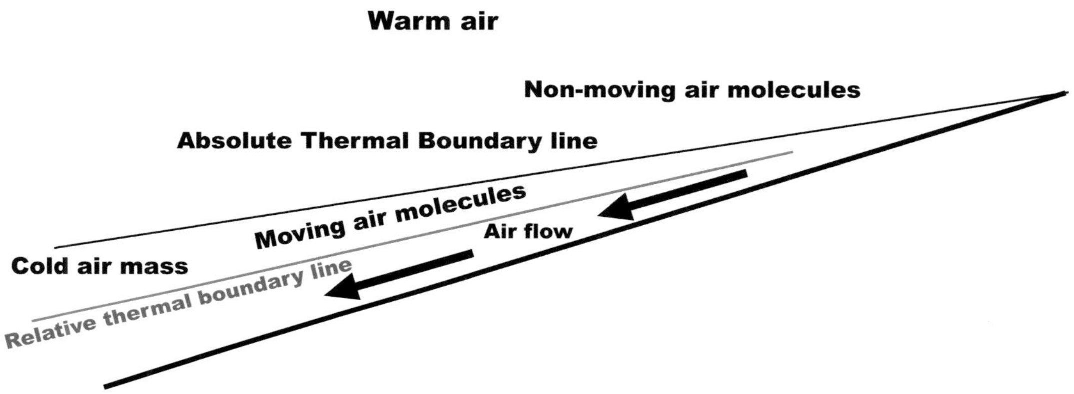

Warm air

Non-moving air molecules

Absolute Thermal Boundary line

Moving air molecules

Air flow

Cold air mass

Relative thermal boundary line

Fig 19 – The thermal boundary line is defined as the separation between moving and non-moving molecules.

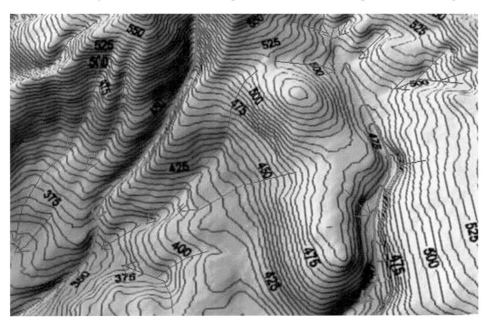

Fig 20 – Streamlines of cold air originate in the smaller gullies and swales and then converge into larger swales.and continue their flow downhill.

63

Fig 21- A cold air river in a swale.

Accumulation Process # 3 – Converging Air Currents

Where two or more dynamic air currents or cold air rivers converge at opposing angles, they can create a pooling of cold air before finally draining down hill. This type of pooling is not dependent on a barrier to air flow, containment, or a lower slope angle. The pooling is caused by chaotic air currents formed by the collision of air streams. At the point of this collision there are backwards (eddy) air currents that are created which will slow the descent of the air streams and cause a buildup. This phenomenon also occurs in gullies and swales that turn sharply causing chaotic air currents and a pooling of cold air at the bend.

Figure 16 shows air currents converging at approximate right angles. The greater the opposition of the angle of the convergence, the more likely that there will be a pooling of cold air at the convergence point.

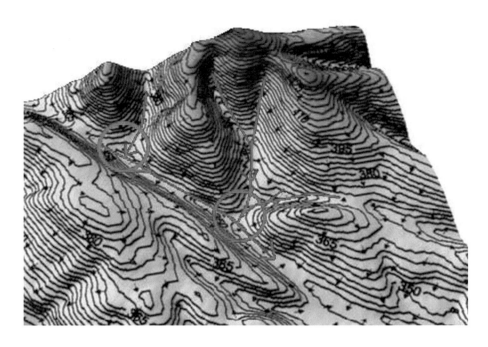

Fig 22 – Opposing convergence of air currents and subsequent pooling areas.

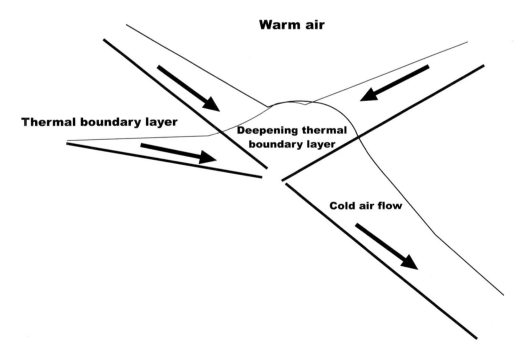

Fig 23 - Converging air currents will cause a buildup of cold air due to chaotic and eddy (backwards) currents. Downstream, the air flows will resume and the thermal boundary layer will lower.

Fig 24 – A sharp bend in a gully can cause a buildup of cold air due to the formation of eddy and chaotic air currents.

Accumulation Process #4 – Lessening Slope

Lessening slope accumulation occurs where cold air that is flowing down hill slows because the slope angle gets less.

This phenomenon is common where crops are planted at the foot of a hillside where the slope begins to flatten. At the point of the lesser slope, the cold air flow slows causing the cold air to accumulate and the thermal boundary line to rise.

Damage up on a hillside can be caused by a variable slope angle where the cold air slows as the slope angle lessens and then picks up speed again where a steeper angle resumes. In the areas of lesser slope, cold air drainage will slow, accumulate and force the thermal boundary layer to rise. Where the relative thermal boundary line submerges the crop, frost damage can occur.

There are 2 main factors that explain the buildup of cold air due to a lessening slope:

1. The amount or mass of cold air that is flowing downhill is determined by heat loss from the surrounding ground that supplies cold air into the crop area and,
2. The speed of the air flow is determined by the angle of slope.

The mass of the cold air that is flowing downhill is determined by heat loss from the ground in the air shed area above. If the cold air mass slows down, but the amount of cold air remains constant there will be a deepening cold air mass as the molecules pile on top of each other.

The down slope flow of cold air is caused by gravity. The angle of the slope, density of the air and the smoothness of the ground surface are the factors that determine the speed of the air flow.

The greater the angle of slope the faster the air will move. The surface of the ground is a factor in the speed of down slope air movement. The smoother the surface, the faster air will move. Hard packed, smooth surfaces will allow more drainage of cold air than disked surfaces or those with cover crop or other obstacles to air drainage.

The cold air mass flowing downhill will slow and build up at the point where the slope lessens and the thermal boundary layer will deepen there. This can cause crops to be submerged in lethal cold air at this point.

If the ground then resumes its prior slope or opens up, cold air will speed up or spread out causing a thinning of the thermal boundary.

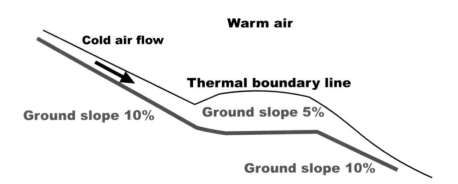

Fig. 25 – Where the slope lessens, cold air accumulation causes the thermal boundary to rise.

The rate of heat loss from the ground is unrelated to the accumulation of cold air down hill in the same way that water flowing off a hillside into a lake is unaffected by the depth of the lake. The sources of water flowing in a stream or river are not affected by accumulation in lakes and ponds downstream.

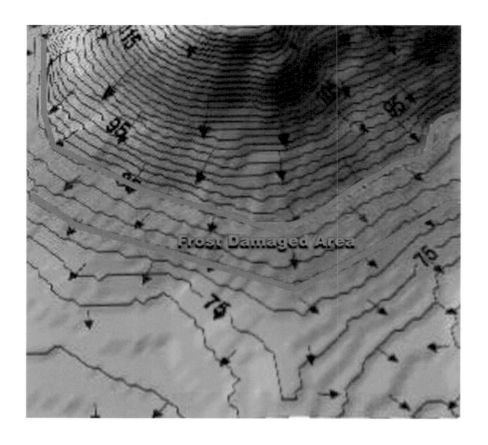

Fig 27 – Damage begins to occur at the point of the lessening slope angle.

Accumulation Process #5 - Plain Slope

A plain slope has a relatively constant angle of drop and little or no converging air currents on the slope.

Accumulation of cold air on a plain slope is caused by the steady deepening of the thermal boundary layer due to the consistent addition of cold air to the slope. The additional cold air is generated from the heat loss of the ground in the slope and orchard area itself, as well as the cold air coming in to the growing area from up hill areas flowing into the crop area. Cold air generated on the slope and outside the growing area, combines causing a deepening of the thermal boundary layer if the slope angle remains constant. Under these conditions, the slope will eventually not be sufficient to drain the cold air fast enough to avoid accumulation.

As in a lessening slope, the mass or volume of cold air moving down the slope is determined by the heat loss from the ground, while the speed of the cold air movement is determined by the angle of slope.

Since the angle of slope on a plain slope is relatively constant, the speed of the air mass moving downhill is also constant. The volume of cold air continues to increase which causes the depth of the thermal boundary layer to increase due to the added volume of cold air coming from the area of the slope itself and outside the growing area.

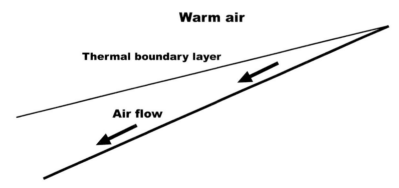

Fig 28 – A deepening thermal boundary layer due to increasing mass of of cold air.

Where the slope angle is steep enough to adequately drain the cold air being generated up hill and from the slope itself fast enough to avoid accumulation that causes the thermal boundary line to submerge the crop, and the slope is higher than the lower accumulation zone, then the area is considered to be 'frost free'.

Where the thermal boundary layer builds high enough to submerge the plant in lethal cold air, then damage will begin at the point where the trees get submerged and increase in intensity further down the hill. A typical damage pattern on a plain slope would be that the trees higher up the hill get frost damage on the lower parts and the damage steadily affects higher up in the trees the further downhill they are.

Fig 29 – A plain slope

Fig 30 – Plain slope where there are no converging air currents and consistent slope angle.

Accumulation Process #6 – Flooding

Flooding occurs when cold air builds up below the growing area and then overflows back into the growing area.

Since the cold air builds up from the basin below the crop area, in a flooding accumulation the area below is colder and gets colder sooner than the crop area. This causes the lower crop rows that are closer to the flooded basin to be more damaged than the rows further inside, even if the elevation is the same. In most instances the basin below will not overflow its normal capacity during every frost event. Most basin areas do have some natural drainage. Where there is some drainage then buildup will occur only when the frost event is severe enough to produce more cold air flowing into the lower basin than the natural drainage can expel. When the buildup of cold air exceeds the holding capacity of the basin, cold air will flood into the growing area.

Flooding occurs when the growing area itself has sufficient drainage capability, but the area below has insufficient drainage and accumulates cold air.

Flooded areas are basically on the 'shoreline' of a cold air river or cold air lake. When cold air that accumulates in the lower basin or streamline rises above the height of the growing area, incursion into the growing area occurs. This can be compared to building a house on a river or lake and the river or lake overflows its banks and 'floods' into the house.

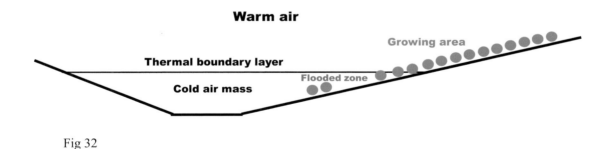

Fig 32

When flooding is caused by a static mass of cold air, such as a cold air lake then the flooded zone will follow the topographic gradient lines the same way a lake will leave a high water mark around its basin.

When flooding is caused by a dynamic mass of cold air such as a cold air river, then the damage zone will cross the gradient lines but be a relatively consistent height above the bottom of the drainage.

In fig 33, cold air generated from the surrounding area is flowing into the basin. Due to the low slope angle of the basin there is insufficient drainage capacity to keep the cold air from building up creating a cold air lake. During periods of severe frost, the cold air level will rise in the basin and flood back into the growing area causing damage along a gradient line.

There are limits to the potential height of the absolute thermal boundary line. The build up of cold air is limited to the thermal boundary created by the highest containment barrier around the basin. Any excess above this level will simply spill over into another area. When the relative thermal boundary line exceeds the absolute thermal boundary line, a regional temperature deficit may exist for the crop trying to be grown.

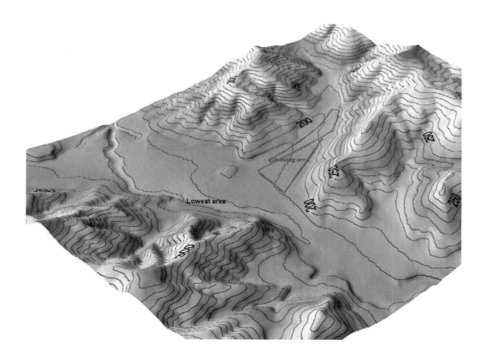

Fig 33 – A growing area that is easily flooded due to insufficient drainage capacity in the main basin.

In a dynamic flow or cold air river, the damage pattern will follow the boundary of the drainage maintaining a consistent distance into the orchard and crossing elevation gradient lines.

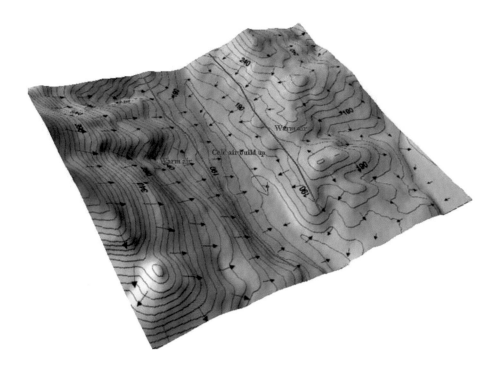

Fig 34 – Damage pattern along a flooded drainage

Fig 35 – A vineyard planted along a drainage canal. The lowest rows will be damaged more severely and earlier than the inside rows because the cold air that causes damage starts to accumulate in the drainage and then floods into the vineyard.

Accumulation Process # 7 – Flat Area

Flat area accumulation is the result of the build up of cold air solely from the area directly underneath the growing area.

In a flat area there is no slope and so there is no movement of cold air. The main driver of the cold air buildup is heat loss from the ground. The ground loses heat through long wave radiation, cools, and then cools the atmosphere from the ground up. Where there is no air movement due to low or no slope and no differences in the soil that would cause differing heat loss rates, cold air will build up evenly along the ground causing a damage pattern that affects the lower levels of the crop first and most severely. The higher up the tree, the less affected it is.

In flat area accumulation there is no direct outside influence of cold air from surrounding higher areas. Some large flat basins that are surrounded by hills can have an overall temperature loss throughout the region. The distance of the surrounding hills and mountains from the growing area is a factor in whether or not the cold air generated on the hillsides will affect the growing area.

Another factor is the length of the frost event. When frost conditions start early in the evening and are not interrupted by wind or clouds, the cold air generated from surrounding hills will be able to travel further. If there is incursion of cold air from outside the growing area that combines with the cold air generated within the growing area, the chance for frost damage increases.

There may be temperature differences in areas of flat topography due to the soil conditions. Sand, loose soil and dry soil will lose heat faster than heavier or wet soil and dark soil will absorb more heat during the day than light soil. During the night, these soils will cool more rapidly, and cool the atmosphere directly above them more rapidly. Because there is no movement of air and thus no mixing, then the air temperature above the sandy areas will be colder.

Fig 36 shows a static cold air mass over the looser, sandy or dry soils in a flat area. These static air masses mimic cold air lake accumulation areas.

Warm air	Cold air	Warm air	Cold air	Warm air
	Sand area		Sand area	

Fig. 36 – the atmosphere directly over soil that has less ability to retain heat will cool faster and cause non contained cold air accumulation that mimic cold air lakes.

Other variations in temperatures over a flat area can be attributed to shading from vegetation, differences in soil moisture content, ground cover variation and soil condition such as disked or hard packed.

Appendix 1 – Evaluation information form

SHuR FARMS®
Frost Protection
Toll Free 1-877-842-9688

1890 N. 8th Street, Colton, CA 92324 (909) 825-2035, (909) 825-2611 Fax

info@shurfarms.com www.shurfarms.com

SYSTEM DESIGN FORM

There is *no charge* for a customized Shur Farms® system recommendation. To understand your frost situation, please provide as much of the following information as possible. You are welcome to attach additional information. Please return via e-mail, fax, or standard mail.

CONTACT INFORMATION

Contact Name:_____ Position: _____ Date: _____

Farm/Company Name:_____

Mailing Address: _____City_____ Zip_____

Physical Address:_____City_____ Zip_____

Phone # _____Cell #_____

Fax #_____E-Mail: _____

FROST AREA DESCRIPTION & HISTORY

Name of Block(s) to be protected: **Area 1**_____ **Area 2**_____

Total Acreage of Block(s): **Area 1**_____ **Area 2**_____

Row Spacing: **Area 1** _____ **Area 2**_____

Crop & Variety: **Area 1**_____ Age: _____

Area 2_____ Age: _____

What is your cordon or tree skirt height? **Area 1**_____ **Area 2**_____

What frost protection method(s) are currently used? (circle)

Area 1:   Wind Machines   Cold Air Drain   Heaters   Under Tree Sprinklers

Over Tree Sprinklers   None   Other(explain)_____

Area 2:   Wind Machines   Cold Air Drain   Heaters   Under Tree Sprinklers

Over Tree Sprinklers   None   Other(explain)_____

Crop loss? (please circle)

Area 1:	less than10%	10 -25%	25- 50%	50%- 75%	more than75%
Area 2:	less than10%	10 -25%	25- 50%	50%- 75%	more than75%

For tree crops, do you have a crop at the very top of the trees? (circle)

Area 1:	Always	Sometimes	Never
Area 2:	Always	Sometimes	Never

Average frequency of frost damage: (circle)

Area 1:	Yearly	Every 2-3 years	3-6 years	6-9 years	10+ years
Area 2:	Yearly	Every 2-3 years	3-6 years	6-9 years	10+ years

What months do you receive frost damage? (circle)

Area 1:	Jan.	Feb.	Mar.	Apr.	May	Sept.	Oct.	Nov.	Dec.
Area 2:	Jan.	Feb.	Mar.	Apr.	May	Sept.	Oct.	Nov.	Dec

What is the minimum temperature during frost season? **Area 1:**_____ **Area 2:**_____

SURROUNDING AREAS

Please circle the locations for what surrounds your growing area.

Area 1:

	North	South	East	West
Your additional growing area	North	South	East	West
A neighbor's crops	North	South	East	West
Residence	North	South	East	West
Out Buildings	North	South	East	West
Forest	North	South	East	West

* Forest below 3 ft -   Light   Med Heavy   Forest above 3 ft - Light Med Heavy

	North	South	East	West
Open Field/Meadow	North	South	East	West
Creek/River	North	South	East	West
Pond/Lake	North	South	East	West
Canal Bank	North	South	East	West
Elevated Road	North	South	East	West
Wind machine closer than 375'	North	South	East	West

Shrubs/Hedges	North	South	East	West

Shrubs below 3 ft -   Light   Med   Heavy   Shrubs above 3 ft -   Light   Med   Heavy

Other obstructions?	North	South	East	West

(Such as compost pile, building, etc.) Please describe_____

Area 2:

Your additional growing area	North	South	East	West
A neighbor's crops	North	South	East	West
Residence	North	South	East	West
Out Buildings	North	South	East	West
Forest	North	South	East	West

   * Forest below 3 ft -   Light   Med   Heavy   Forest above 3 ft -   Light   Med   Heavy

Open Field/Meadow	North	South	East	West
Creek/River	North	South	East	West
Pond/Lake	North	South	East	West
Canal Bank	North	South	East	West
Elevated Road	North	South	East	West
Wind machine closer than 375'	North	South	East	West
Shrubs/Hedges	North	South	East	West

Shrubs below 3 ft -   Light   Med   Heavy   Shrubs above 3 ft -   Light   Med   Heavy

Other obstructions?	North	South	East	West

(Such as compost pile, building, etc.) Please describe_____

BLOCK MAP

If available, please attach a scale block map, Google earth image, or a hand drawn map of the block(s) to be analyzed. On the block map mark off:

1) Boundary lengths and/or acreage
2) Row Direction
3) Locations & methods of currently used frost protection (wind machines, sprinklers, etc.)
4) Buildings, roads, canals, forests, ponds, fields, etc.
5) Mark the frost damage areas in different patterns/colors to show serious, intermediate, & minor damage. Please be as specific as possible.

Summary Vineyard Frost Analysis

Introduction
Subject vineyard is located at 999 Stone Dr., Lake County, Cal.. USGS Lakeport (CA) Topo map (fig. 1)

Year planted 2001. Approximately 9 acres.

Cordon Height 42".

History
This vineyard has been planted since 2001 and has not experienced frost damage. In addition, similar vineyards planted with the same regional and micro climate zones have also not experienced frost damage in the same period.

Frost Occurrence Summary
Radiation frost is caused by ground cooling due to heat loss via long wave radiation. The ground loses heat and cools, and in turn cools the atmosphere. The colder air molecules are more dense and heavier than the warmer ones. The coldest air is nearer to the ground and as height from the soil increases, the temperature increases. When plant tissue (cordon) is submerged in lethal temperatures for a sufficient period of time, frost damage occurs.

Most often, this situation is caused by an accumulation of cold air due to katabatic air flows. The heaviest (coldest) air molecules are affected by gravity and will flow down slope if allowed to do so with out obstruction. Another cause of accumulation is converging air currents as in a swale.

Also, a trait that most frost prone areas have in common is that they are in wind protected areas.

Subject vineyard is flat and none of these conditions are present here.

Recommendations
There are two general types of frost protection – *Active* measures and *Passive* measures. Active measures are any type of protection that requires power or Btu input such as water, conventional wind machine (air mixer type), heaters, and Shur Farms Cold Air Drain® method.

The above active measures are not suitable for all types of terrain or problems. Some methods can cause extreme damage if not used properly or if used in an inappropriate area.

Over vine water, whether micro sprinklers or solid set sprinklers should never be used outside wind protected areas as wind will enhance evaporation and chill the plant tissues. In some cases, this chilling is far below what would have occurred if no water was used at all. <u>We do not recommend this method in this vineyard due the high probability of failure and subsequent catastrophic damage.</u>

Subject vineyard is not in a wind protected area, and strong winds may appear suddenly and with out warning.

<u>Conventional wind machines</u> (wind mixers) are best suited to flat topography such as the subject vineyard, but are not with out risk. It is common during a frost night that clouds or wind will appear. When this happens, the atmosphere turns from radiation, or stratified atmosphere, to advection. Advection is characterized by colder air as the height from the ground increases, not warmer air as in radiation frost. This change can happen suddenly and with out warning. When it does, it is usually sufficient to simply turn off the wind machine to avoid blowing colder air into the vineyard and causing damage. It would be necessary to install 1 – conventional WM for this vineyard. Energy usage for this method is approx. 15 gallons per hour of gas, diesel or propane.

<u>Heaters</u> will provide benefit in areas that have no katabatic air movement such as in the subject vineyard. In areas of dynamic air flows (down slopes and swale areas) heaters are generally useless as the cold air is continually being replaced. Heaters are expensive to run, generally burning about 40 gallons of fossil fuel per acre/hr. A six hour night using heaters at a cost of $2.50 per gallon would cost $600.00 *per acre per night.* The obvious negative to this is the high cost and the reluctance of a grower to actually turn on his frost protection.

Smoke from heaters or 'smudge pots' is not beneficial due to the size of the individual particles –about 1 micron diameter. This is far smaller than the long wave radiation that is transmitting heat from the ground, but sunlight that warms the ground is short wave radiation. Smoke in the air will inhibit the warming of the soil when the sun comes up.

<u>Under vine irrigation</u> is a safe method, although limited in its effects. Because the water is either frozen at, or absorbed into the ground, there is no danger of wind chill or evaporative cooling if the wind comes up. The effect of under vine irrigation is directly dependent upon the volume of water being applied, and how much of the water is converted into ice. It is necessary to have the correct spray pattern for the volume and pressure of water available in order to convert as much water to ice before the water soaks into the ground as possible. Normal rates of application are usually 25 – 35 gpm per acre. The main negative with this method is the limitation of a particular soil type to adequately absorb and drain off excess moisture. The soil should be analyzed prior to installing this method to avoid water caused diseases in the vines.

Cold air drainage can be accomplished by extraction of the heaviest (coldest) air molecules. As the colder and heavier molecules along the ground are drawn into the machine, the warmer air molecules above come closer to the ground to replace them. The cold air entering into the machine is then propelled approximately 300 ft into the air and drained into the upper inversion layer. Because the Cold Air Drains® only recognize physical boundaries, with this method it is necessary to surround the vineyard with a barrier 3 – 4 ft high or higher to prevent cold air from the surrounding area from entering into the protected area. The barrier can be natural or artificial such as soil berms, heavy vegetation, or plastic sheeting. In the subject vineyard, it the installation of one Shur Farms Cold Air Drain® model 1550 is recommended. Fuel usage is approximately 1 GPH.

Passive Measures

Passive measures are essential in conjunction with an active frost protection system in this vineyard. In the subject vineyard we recommend the following passive measures-

Cover crop management in the rows. In order to maximize the absorption of heat from sunlight during the day into the ground, it is recommended to keep cover crops in the rows to no higher than 2 ½". Mowing (or chemical spray) is preferable to disking.

Cordon height management. Cordons should be maintained at the highest possible level from the ground. The air is warmer higher from the ground. A similar vineyard planted just south of the subject vineyard has cordons set at 36", while the subject vineyard is at 42". Since the similar vineyard also has not experienced frost damage in at least the last 6 years, this would indicate that the subject vineyard would be able to withstand even colder events than that which have occurred.

Soil moisture. To maximize heat retention and also heat absorption into the soil it is recommended that the soil be kept as moist as possible in the first 12" of depth. Keeping the soil dark will increase absorption of heat and higher moisture levels will tend to help retain this heat through the night.

Conclusions

While the subject vineyard is not considered particularly frost prone, there can be substantial improvement to lessen frost risk using passive measures.

Considering the relative low level of frost risk, it is advisable to disregard the usage of any active method that has the potential of causing damage. The risk –benefit relationship here is not sufficient to compensate for the possible losses.

It is reasonable to protect with passive measures to achieve a decrease in frost risk commensurate with overall risk in the industry. If additional protection is desired, it is advisable to use one of the safe methods outlined - <u>Under vine</u> water, <u>heaters</u> or <u>Cold Air Drain</u> method, or any combination of these.

Fig 1 - Shows subject vineyard overlay on regional and DEM. Contour lines 1M each.
    (Source of aerial photo Google Earth.)

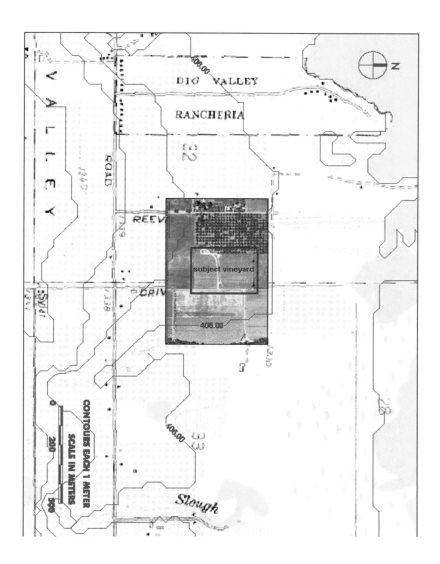

Appendix 3 –Sample detailed topographic, air flow and optimization plan summary analysis (Part 2):

SHuR FARMS®

Frost Protection

Division Recovery P.T. Corporation

1890 N. 8th Street, Colton, CA 92324   (877)-842-9688, (909) 825-2611 Fax

info@shurfarms.com

www.shurfarms.com

Frost Protection Proposal for:

AraVineyards

Fig. 1 Surface Vector 1 meter DEM
Topographical and Airflow Analysis

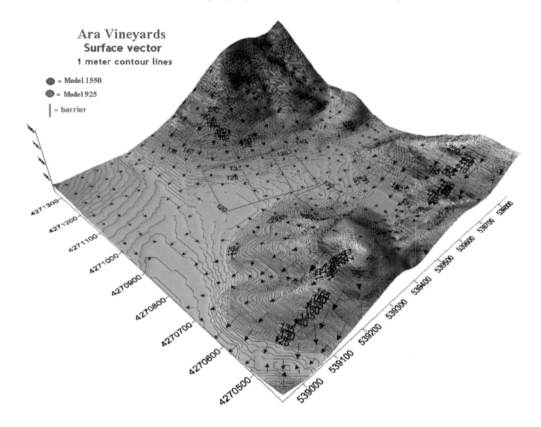

This proposal is based on some or all of the following, as well as other agronomical information provided by alternate sources:

1.   Information supplied by the grower
2.   Information collated onsite by a representative for Shur Farms® Frost Protection or an authorized dealer.
3.   USGS topographical information
4.   USDA agronomical information
5.   State, county and local agriculture resources

General Characteristics

The vineyard is located on USGS Calistoga (CA) quad 7.5 min.

Cause of Frost Damage in the Vineyard

Block 7 – Cold air accumulation due to converging cold air streamlines and insufficient drainage in the basin.

Block 9C- Cold air is accumulating in the block due to a lessening slope that is causing air movement to slow and cold air mass to deepen.

Blocks 10 & 11 – Cold air accumulation due to a lessening slope, causing air movement to slow and cold air mass to deepen.

Fig. 2 – Looking north from the top of blk 7. Shows the low

Fig. 3 looking southwest into blk 7. area tBasin area.

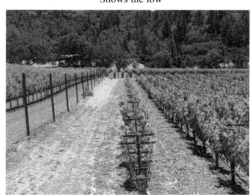

Fig. 4- Block – looking east into damaged area. Shows lessening slope to the south.

Fig. 5 - block 10 & 11 looking west along the north boundary. Shows the vegetative barrier along the boundary.

System Recommendation

Block 7

1. Quant. 1 – Shur Farms Cold Air Drain® Model #925 located in the lowest or most convenient spot nearest the recommended location in the basin at approximately row 25.
2. Install barrier 5ft high x 120 ft long 15 -20ft downstream of the machine and perpendicular to the slope.
3. Maintain cover crop at 3" or less inside the vineyard.
4. Maintain and encourage the natural vegetative barriers along the west, south and east boundaries of the block.

Block 9

1. Quant. 1 – Shur Farms Cold Air Drain® Model #925 located in the damage area at the top portion of the block, in the lowest or most convenient spot nearest the recommended location.
2. Install barrier 5ft high x 90 ft long 15-20ft downstream the machine.
3. Maintain cover crop at 3" or less inside the vineyard.

Block 10 & 11

1. Quant. 1 – Shur Farms Cold Air Drain® Model #1550 located in the damage area at the top portion of the block, in the lowest or most convenient spot nearest the recommended location on the road.
2. Install barrier 5ft high x 165 ft long 15-20ft downstream the machine, positioned to extend 75 ft into block 11 and 90 ft into block 10.
3. Maintain and encourage the existing vegetation along the north boundary of blocks 10 and 11. **Do not add to the vegetative barrier here as it may affect the vineyard upslope.**
4. Maintain cover crop at 3" or less inside the vineyard.

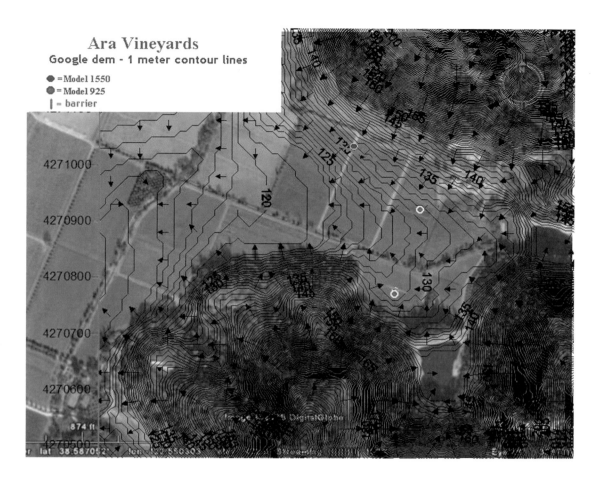

Fig 4 –Machine and barrier locations.

Printed in the United States
By Bookmasters